**Principles of Microbiology**

Withdrawn

# Principles of Microbiology for students of food technology

## Second edition

**Thelma J. Parry and Rosa K. Pawsey**

**Hutchinson**
London   Melbourne   Sydney   Auckland   Johannesburg

Hutchinson & Co. (Publishers) Ltd

An imprint of the Hutchinson Publishing Group

17–21 Conway Street, London W1P 6JD

Hutchinson Group (Australia) Pty Ltd
30–32 Cremorne Street, Richmond South, Victoria 3121
PO Box 151, Broadway, New South Wales 2007

Hutchinson Group (NZ) Ltd
32–34 View Road, PO Box 40–086, Glenfield, Auckland 10

Hutchinson Group (SA) (Pty) Ltd
PO Box 337, Bergvlei 2012, South Africa

First published 1973
Reprinted 1979, 1981 (twice), 1982
Second edition 1984

Set in times by Activity Ltd, Salisbury, Wilts.


Printed and bound in Great Britain by
Anchor Brendon Ltd, Tiptree, Essex

**British Library Cataloguing in Publication Data**
Parry, T.
    Principles of microbiology.
    1. Food – Microbiology
    I. Title     II. Pawsey, R.
    576′.163     QR115
ISBN 0 09 152561 6

# Contents

# Preface to the first edition

Our aim is to introduce food microbiology in such a way that the student will gain a basic grounding in this subject. The student will then be in a position to understand more authoritative texts on the subject if this proves necessary.

We have directed this book towards the needs of those students who possess no formal knowledge of microbiology but are pursuing post Ordinary-level courses in the various branches of the food industry.

We anticipate that the book will be suitable for students following courses in food technology, catering, RSH and the various bakery technology subjects – all of which require a knowledge of microbiology. We also think the text suitable for the requirements of dietitians and home economics teachers in training.

*Thelma J. Parry and Rosa K. Pawsey 1973*

# Preface to the second edition

The original text has been rearranged to a certain extent, and expanded. Two entirely new chapters on water in food operations and quality control have been added. The chapter on pest infestations has been removed from the main text to make it more distant from the microbiology text, but it is retained to form a second appendix in recognition that this subject area is often included in 'hygiene' courses. In the ten year period since the publication of the first edition, courses have changed both in content and style, but the aim of this edition remains the same – to provide a basic grounding in food microbiology for those students whose careers will in some way be bound up with the production, sale or service of food.

*Rosa K. Pawsey 1983*

# Acknowledgements

I would like to express my thanks to Thelma Parry for her help and support, for her contributions and for her constructive criticism of the draft of this edition.

I would also like to thank Mr. T. C. Tamplin of APV International Ltd for his help with Chapter 10, particularly with regard to CIP operations and systems. I must also thank APV International Ltd for providing the photographs for Figures 70 and 71.

I am most grateful to Dr Richard Harding of MAFF who gave me information and guidance on EEC legislation, and Mr A. Murphy, Environmental Health Department, Cardiff City Council who checked Appendix 2 and made suggestions to bring it up to date.

Finally, I would again like to thank Mr Edward Meyrick, now in the PHLS at the Central Public Health Laboratory, Colindale, London, who has supplied the photographs for Figures 1, 2, 3, 6, 15, 20, 22, 24, 54 and 69, for which I am most grateful.

# Introduction

The presence of micro-organisms in the environment around us is so common that it rarely attracts notice. We are so used to the fact that milk left in the sun will 'go off' and that vegetable material in a compost heap will 'rot' that we give little thought to the underlying causes of such changes. Both these processes and many others are brought about by the growth and activity of micro-organisms whose existence has been known to man for about three hundred years.

Serious microbiological research was initiated by a Dutchman, Antonij van Leewenhoek (1632–1723), who ground lenses of sufficiently good quality to enable him to examine small drops of water and other material under the microscopes he made. In so doing he noticed the presence of small organisms moving about in the liquids. Interesting though his discovery was, it was treated by his contemporaries merely as a scientific curiosity and not as significant to man's well being. The task of showing this remained for the scientists of the nineteenth century.

There were a number of eminent workers in this field foremost among whom was Louis Pasteur (1822–1895), now recognized as one of the founders of the science of microbiology. He studied fermentation and demonstrated that it was the growth of yeasts and bacteria which caused wine to ferment or sour when it was bottled or casked. He developed a process of heating wine to 50 to 60°C to kill the organisms which caused the spoilage without altering the quality of the wine. In so doing he rendered a great service to the French wine industry which had been losing trade due to its products souring during export. The process is now widely applied and is known as *pasteurization*.

Joseph Lister (1827–1912) working in Scotland heard of Pasteur's work and applied the knowledge to surgery. He realized that many people died because they were infected with harmful bacteria during and after operations. He pioneered antiseptic surgery (destruction of germs) by applying neat phenol (carbolic acid) to wounds. The phenol was effective but rather damaging to the skin, and Lister spent many years improving the technique. He attempted to provide a germ-free atmosphere in the operating room by spraying the room with a 5 per cent aqueous solution of phenol. Eventually the irritating effect of the spray led to its abandonment and was replaced by the technique of asepsis – scrupulous cleanliness.

Robert Koch (1843–1910) demonstrated that anthrax, the fatal disease of sheep and cattle is caused by a bacterium *Bacillus anthracis*. Scientific contemporaries of Koch made many other discoveries in the field of microbiology. Once the importance of micro-organisms had been realized, and with the advances in scientific methods, there was little to hold back research into their activities.

During the twentieth century, great strides have been made in understanding the causes of diseases in man, animals and plants which have led to the control and eradication of some diseases. Bovine tuberculosis – TB contracted from cattle – was rampant in the nineteenth century in the United Kingdom but with the pasteurization of market milk, tuberculin tested herds, and the discovery of drugs such as streptomycin, together with better nutritional standards and better housing, this disease has today been largely eradicated. Other food borne diseases which have been brought under control in this country are diptheria and scarlet fever (although other streptococcal infections are commonly occurring).

In addition to causing disease in man and animals, micro-organisms cause plant and animal material to break down – 'rot' – an activity which has both advantages and disadvantages. It is to our benefit that micro-organisms break-down dead vegetation and 'garbage' so that the elements of which they are composed are returned to the soil. However it is disadvantageous that stored food should be broken down – spoilt, for changes in appearance and flavour lead to its wide scale wastage.

In some manufacturing industries micro-organisms are used at certain stages. Their activities are essential in the formation of foods such as bread, beer, wine and pickles. Some antibiotics are produced as a result of microbial activity and vitamins are extracted from some species grown on a large scale. It is also the case that new types of food made primarily from concentrated micro-organisms which have been grown on industrial by-products and wastes are now on the market. Certain strains of yeasts, for example, can be grown either on heavy fuel oil or pure normal paraffins. In feeding trials, animals have been given feeds substituted with these yeast concentrates. It has been found that the animals accept the food and do not appear to tire of it, and in addition show growth rates comparable to those of animals on normal feeds. Conversion of microbial concentrates into a form acceptable to human beings can be achieved by feeding them to animals which themselves are destined to be human food. Biotechnology – the production of food and other materials by exploiting microbial activities – is the fastest growing area of food production technology.

# The importance of micro-organisms

**Advantages**

Micro-organisms are involved in:

1  Decomposition of organic material – 'compost'.
2  Fermentation of processes in the manufacture of foods such as bread, beer, wine and pickles.
3  Manufacture of vitamins; for example proprietary brands of yeast extract are high in vitamins of the B complex.
4  Potential uses as food or as food supplements.

**Disadvantages**

Micro-organisms cause:

1  Disease in people and animals, for example food poisoning, the food and water borne diseases of typhoid and dysentery.
2  Spoilage of food.
3  Hold-ups in industrial processes by growing in and blocking pipes as, for example, in sugar refining.

# Chapter 1

# Introduction to micro-organisms

Micro-organisms are not all alike but they share a common feature in that for the most part they are individually invisible to the naked eye and can only be seen when magnified. Size alone determines which organisms are included in this class.

Micro-organisms differ from one another in appearance and activity. The types to be discussed in this book may be divided into six major groups:

1 Protozoa
2 Algae
3 Viruses
4 Bacteria
5 Yeasts
6 Moulds

## Protozoa

These are simple unicellular animals of which there are many known species. They live in an aqueous environment such as pond or ditch water, sea or soil water. The majority are free living and harmless to man, a well known example being the amoeba. A few species of protozoa are of considerable importance because they cause diseases such as malaria, sleeping sickness and amoebic dysentery in man and animals. Amoebic dysentery is a water and food borne disease caused by *Entamoeba hystolytica*, a pathogenic protozoan which fortunately is seldom encountered in this country but which is, on occasion, imported by people from overseas.

## Algae

This is a group of simply constructed plants, some of which are large (macroscopic), for example the large types of seaweed. Others are very tiny (microscopic) and are only visible under the microscope. All algae manufacture their own food by the process of photosynthesis (the manufacture of carbohydrate by the combination of water and carbon dioxide in the presence of chlorophyll using light as a source of energy). The microscopic algae are usually free living organisms found where there is water and sunlight available to them. They are commonly seen as green slime on the surface of ponds and aquaria.

## Viruses

Viruses are the smallest of all micro-organisms varying from 10 nm to 300 nm diameter*. Viruses can only be seen when viewed under the electron microscope which gives magnifications in excess of 25,000 diameters. Viruses are themselves metabolically inert, but they can enter living cells and redirect the activities of the cells towards replicating themselves. This process of multiplication causes the death of the infected cells and results in disease of the host organism. In this process the viruses undergo

---

* nanometre (nm) $= 10^{-9}$ metre
  micrometre (μm) $= 10^{-6}$ metre

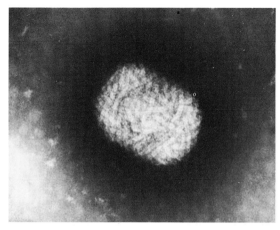

(a)  Molluscum contagiosum (× 185,000)

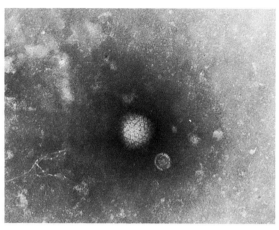

(b)  Herpes simplex (× 380,000)

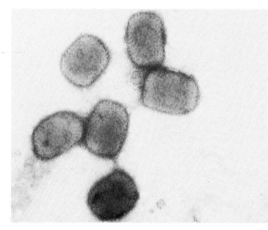

(c)  Vaccinia (× 120,000)

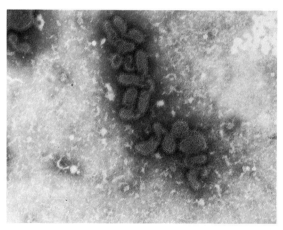

(d)  Hong Kong flu (× 160,000)

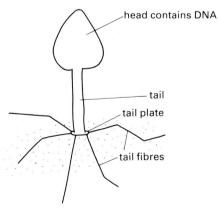

(e)  Bacteriophage

Figure 1    *Viruses*

conversion into non-infective forms as a necessary step in their multiplication.

The simplest viruses are comprised of only nucleic acid wrapped in a protein coat, but a wide variety of shapes, sizes and complexity are known to exist – features primarily revealed through the use of the electron microscope.

Some viruses, known as *bacteriophages* ('phages), specifically attack bacteria by binding to the bacterial surface. Bacteriophage attack in lactic acid starter bacteria adversely affects the starter activity and is a considerable problem in the dairy industry (Figure 1).

Originally, the viruses tended to be classified according to the tissues that they invaded, or on the more easily observed features, such as their host range; but a newer and better system of classification based on their unchanging features is currently being worked out. It will, however, be some time before the newer system entirely replaces the old. Examples of the older names of groups of viruses followed by the appropriate newer name are given here:

Enteroviruses – native to the intestines of man
 and animals (Picornavirudae – includes the
 enteroviruses)
Adenoviruses (Adenoviridae)
Pox viruses (Poxviridae)
Arbo viruses (included in the Togaviridae)

Examples of animal diseases caused by viruses are poliomyelitis, measles, and smallpox; and of plant diseases, tobacco mosaic disease. Viruses other than bacteriophages do not appear to be of as great a significance to the food industry as bacteria, yeasts and moulds, but some viruses are known to be transmitted by contaminated food and water. Infectious hepatitis, for example, has been known to be transmitted in this way (see page 80). Improvement in isolation techniques, including the routine use of the electron microscope, is helping in the study of the transmission of virus particles in foods.

## Bacteria

### *Occurrence*

These are simple single celled organisms which

occur widely. The body fluids of warm blooded animals, including man, will support their growth, for example in septic cuts and the moist passages of the nose, mouth and throat. The armpits, the groin, the umbilicus, the bowel and the lower urinary tracts are other areas inhabited by bacteria. Bacteria also occur in the soil, on plants, in air and in water. In fact, very few places, if investigated, fail to reveal their presence.

### *Morphology*

Bacteria are of simple shape and can only be seen individually with the aid of the microscope. Their shape may be:

1  Spherical – coccus (plural – cocci)
2  Short plump rod – cocco-bacillus (plural –
   cocco-bacilli)
3  Rod shaped – bacillus (plural bacilli)
4  Spiral – spirillum (plural spirilla)
5  Comma shaped – vibrio (plural vibrios)

When stained films of bacteria are looked at under the microscope many bacterial cells are seen at the same time, and it is sometimes a means of making a preliminary identification if the arrangement of the cells is observed (see Figures 2, 3 and 4).

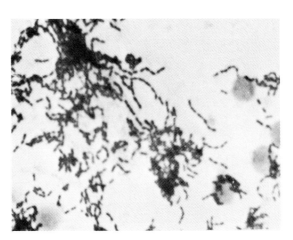

Figure 2  *Streptococci* (× *1000*)

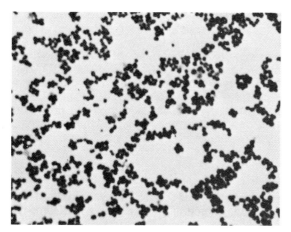

Figure 3   *Staphylococci* (× 1000)

### Structure

Regardless of their individual cell shape all bacteria have the same internal structures. The cell consists of *cytoplasm* throughout which small units, the *ribosomes* (the sites of protein synthesis), and diffuse areas of staining (the nuclear material of DNA and RNA), exist attached to the membrane system. The DNA is a single, very large and highly complex molecule which is divided into regions (genes). These genes have a very specific chemical form and they encode for the specific nature of the organism. The *cell membrane* controls the entry and exit of all substances, and surrounds and

(a)   *coccus* – spherical or near spherical cell. Size varies, but is in the region of 0.5–1.0 μm in diameter

staphylococci – cells arranged in clusters. About 0.5–1.5 μm in diameter
for example *Staphylococcus aureus*
            *Staphylococcus albus*

streptococci – cells arranged in chains. About 2 μm in diameter
for example *Streptococcus faecalis*
            *Streptococcus pyogenes*

diplococci – mainly in pairs. Less than 2 μm in diameter.
for example *Streptococcus pneumoniae*

sarcinae – in packets of eight. About 1.8–3.0 μm in diameter.
for example *Sarcina maxima*

(b)   *bacillus* – rod-shaped cell, longer than it is broad. Size in the region of 0.3–2.2 μm by 1.3–14 μm

small with rounded ends. About 1.1–1.5 μm by 2.0–6.0 μm
for example *Escherichia coli*

large, square-ended. About 1.2–1.5 μm by 2–5 μm
for example *Bacillus megaterium*

(c)   *vibrio* – comma-shaped short rods with a curved axis. Size in the region of 0.5 μm by 1.5–3.0 μm
for example *Vibrio parahaemolyticus*

(d)   *spirillum* – spirally-shaped curved rods. Size in the region of 0.2–0.8 μm by 0.5–5 μm
for example *Campylobacter fetus*

Figure 4   *Shapes of bacteria*

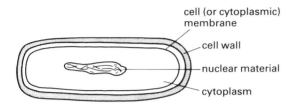

Figure 5 *Simplified structure of the bacterial cell showing features common to all bacteria*

forms structures within the cytoplasm where all metabolic activities take place. The *mesosome*, which is found in many gram positive bacteria, is a membraneous feature which is probably involved in cross-wall formation – important at cell division. The outer cell layer is the mechanically strong *cell wall* of trellis-like structure whose function is to retain the characteristic shape of the organism, and acts as a barrier to certain compounds. If the cell wall is experimentally removed the cell does not disintegrate but becomes spherical (see Figure 5).

The gram's staining technique is important in differentiating bacteria into two groups – gram positive cells and gram negative cells. The test relies on fundamental differences in cell wall

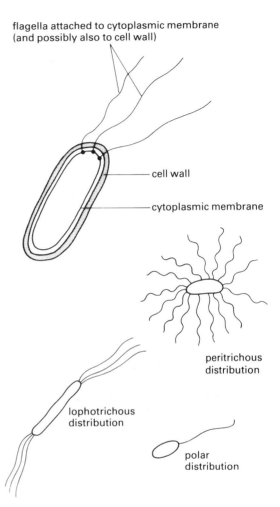

Figure 7 *Flagella – attachment and distribution*

biochemistry and morphology. It is now known that the cell walls of gram negative bacteria are more complex than those of gram positive organisms.

Bacteria sometimes possess structures additional to those already mentioned. A flagellum is a whip-like appendage attached to the cell membrane by a complex basal structure which rotates it. Flagella may be present in groups or singly, or covering the whole of the outside of the cell. They are capable of movement and they are responsible for motility in bacteria (see Figures 6 and 7 and Table 1).

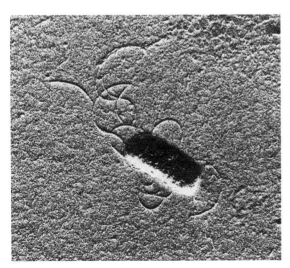

Figure 6 *Salmonella showing flagella (× 55,000)*

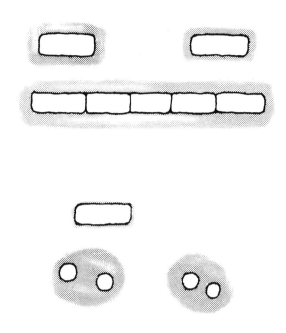

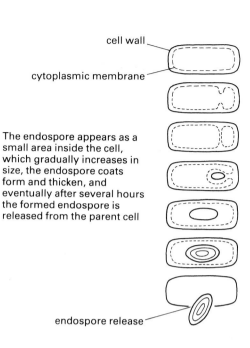

cell wall

cytoplasmic membrane

The endospore appears as a small area inside the cell, which gradually increases in size, the endospore coats form and thicken, and eventually after several hours the formed endospore is released from the parent cell

endospore release

Figure 8   *Capsules – they can be formed by* Diplococcus pneumoniae, Leuconostoc mesenteroides, Bacillus anthracis *and* Bacillus megaterium *and others. The size of the capsule formed depends on the species and varies with the growth conditions*

Figure 9   *How a bacterial endospore is formed. The position of the spore within the cell is characteristic of the species, and can be terminal, subterminal or central. It may also cause the cell wall to bulge, or may be small enough to cause no cell distortion*

Table 1 and Figure 8 indicate some bacteria in possession of a *capsule*. This is a layer of gelatinous material produced by the bacterial cell itself which adheres to the outside of the cell. The capsule may be composed of either complex polysaccharides or of polypeptides. The function of the capsule is not really known, but in some circumstances it seems to protect the bacterium against destruction. The chemical composition of the capsule is often mosaic, yet unique to the species or strain. This fact can be made use of in the serological identification of capsulate organisms (see page 142).

*Endospores*, more commonly referred to as spores, are structures produced by the groups of bacteria genus *Bacillus* and genus *Clostridium*. Other groups of bacteria also produce endospores although their classification is a matter of debate. Endospores are produced singly within cells in response to certain internal and external

Table 1   *Structures possessed by some bacteria*

| Structure | Possessed by |
|---|---|
| Flagella | Some strains of genus *Salmonella* <br> Some members of genus *Bacillus* <br> *Escherichia coli* <br> *Proteus vulgaris* <br> (and other bacteria) |
| Capsule | *Leuconostoc mesenteroides* <br> *Streptococcus pneumoniae* <br> *Bacillus subtilis* <br> (and other bacteria) |
| Endospores | Genus *Bacillus* <br> Genus *Clostridium* |

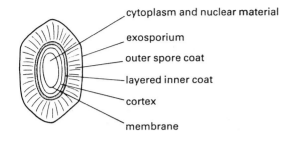

(a) dormant endospore

(b) the endospore is now surrounded by environmental conditions which *activate* it

(c) following activation the spore coat takes in water, specific chemical changes occur and the spore swells. The process is *germination*

(d) eventually the spore coat splits and one vegetative cell emerges. This is *outgrowth*

(e) the vegetative cell completely emerges and leaves the empty spore coat behind

(f) the vegetative cell grows and eventually divides

Figure 10 *A bacterial endospore and the production of a vegetative cell from a dormant endospore*

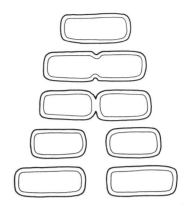

Figure 11 *Binary fission in a bacterium. The cell elongates by growing, the cell wall grows across the cell and divides it into two small cells. Each of these cells grows and will in turn divide*

ing) and to chemicals (such as disinfectants and sterilents). A spore can survive in dust, on vegetation and in soil for weeks, months or even years, or until it eventually finds itself in an environment suitable for reproducing the vegetative cell. For that to happen the following steps are involved. Firstly, *activation*, which requires heat or ageing of the spore, secondly, *germination* and thirdly, *outgrowth*, which releases one vegetative cell capable of actively metabolizing, increasing in size (growing) and of reproducing. (See Figures 9 and 10.)

stimuli. The formed spore is multilayered, comprising the external exosporium, the spore coat and the thick inner cortex. The position of the spore in the cell is characteristic of the species and can be terminal, subterminal or central. It may also cause the cell wall to bulge, or may be small enough to cause no cell distortion. The spore, containing sufficient materials for a new cell to emerge from the spore coat, is released in due course from the parent cell. A mature spore can exist in a dormant state for a long period, being resistant to the adverse effects of severe heat (such as cooking), cold (such as refrigeration and freez-

### Reproduction

Bacteria reproduce by a process known as *binary fission*, (Figure 11), a process of one cell dividing into two parts. This process can lead to a rapid increase in cell numbers, as indicated in Figure 12. Large numbers of bacteria grouped together form colonies which, when the numbers have reached several million cells, may be visible to the naked eye as a very small pin head colony. Species forming larger colonies may be identified to some extent if the colonies develop into characteristic shapes. Figure 13 shows how bacterial colonies form, and Figures 14 and 15 indicate some colonial morphologies.

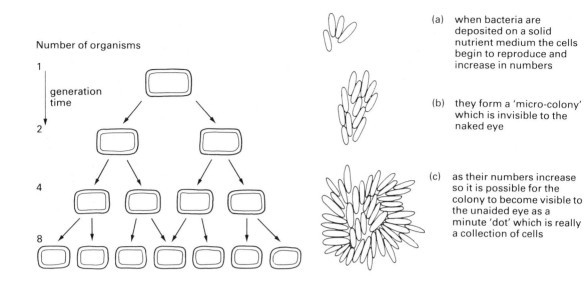

Number of organisms

(a)   when bacteria are deposited on a solid nutrient medium the cells begin to reproduce and increase in numbers

(b)   they form a 'micro-colony' which is invisible to the naked eye

(c)   as their numbers increase so it is possible for the colony to become visible to the unaided eye as a minute 'dot' which is really a collection of cells

Figure 12   *Cell multiplication. The total number doubles when all the cells divide. Generation time may be as short as 10 or 12 minutes*

Figure 13   *How bacterial colonies form*

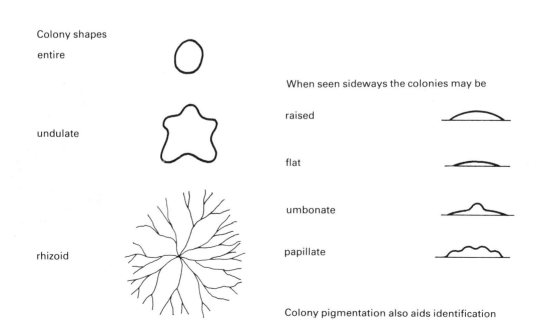

Colony shapes

entire

undulate

rhizoid

When seen sideways the colonies may be

raised

flat

umbonate

papillate

Colony pigmentation also aids identification

Figure 14   *Bacterial colonial morphology*

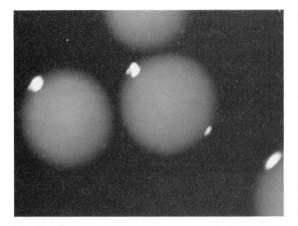

(a)  *E. coli*

(b)  *Bacillus cereus*

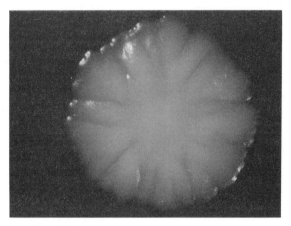

(c)  *Clostridium renale*

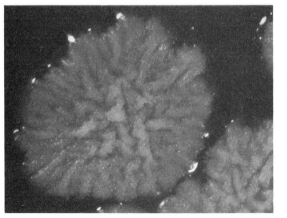

(d)  *Bacillus macerans*

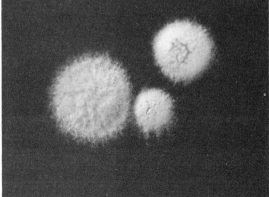

(e)  *Nocardia asteroides*

Figure 15  *Individual bacterial species form colonies which are characteristic of them and which can be used as features for identification in mixed bacterial growth. Some of their shapes are shown in these photographs*

# Yeasts

## Occurrence

Yeasts are single celled organisms which can only be seen individually using a microscope, although a large mass of yeast cells can easily be seen with the naked eye. Yeasts are mainly saprophytic, occurring on the leaves, flowers and on the exudates of plants. They are transported from plant to plant by insects which act as vectors. Yeasts occur in soils but tend not to thrive there – their population being replenished by yeast laden rotting fruits and leaves. Salt water may be populated by some species – the numbers being higher in the presence of organic matter. Little is known about yeast populations of fresh water. The skin and alimentary canal of warm blooded animals may carry saprophytic yeasts. A few species are pathogenic – causing skin infections in man, others cause disease in plants.

Yeasts are used in several industrial processes – in the making of beer, wines, spirits and bread – utilizing their ability to ferment sugars, to produce carbon dioxide and alcohol.

## Morphology

Yeast cells are usually larger than bacterial cells but are nevertheless measured in micrometres*. Most yeast cells have a simple morphology, being oval or rod shaped as shown in Figure 16.

## Structure

The yeast cell has a simple structure. The outer wall is composed of complex polysaccharide, and lying beneath is the cell membrane – both having the same functions as in bacteria. The cytoplasm contains the discrete nucleus, and often a fluid filled vacuole in which the storage products of fat droplets or glycogen granules can occur.

## Reproduction

When yeasts are growing well they reproduce by a process known as *budding* (Figure 17). A small

(a) oval-shaped yeasts for example *Saccharomyces cerevisiae* about 3–7 μm by 4–14 μm

*Saccharomyces carlsbergensis* about 5–10 μm by 5–15 μm

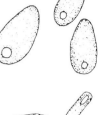

(b) rod-shaped yeast for example *Candida* yeast with elongated cells about 3–5 μm by 5–9 μm

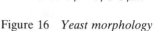

Figure 16   *Yeast morphology*

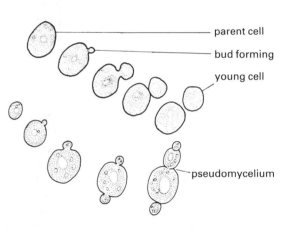

— parent cell
— bud forming
young cell
pseudomycelium

Figure 17   *Budding yeasts – asexual reproduction*

bulge appears on the side of the mother cell which gradually increases in size; at the same time the cell wall of the mother cell constricts, progressively cutting off the daughter cell. When the daughter cell is about half the size of the mother cell the cutting off process is complete and the smaller one is released, which in turn will increase in size until it too is ready to reproduce by budding. Sometimes if a cell is thriving in a medium it will reproduce at a very fast rate and a second bud begin to form before the first is released leading to chain of cells known as *pseudomycelium*. Scars are left on the cell surface where the daughter cell has detached itself (Figure 17). New buds do not form on the scar sites. Some yeasts can also reproduce by a sexual method involving the mating of two cells.

---

* $10^6$ micrometres = 1 metre

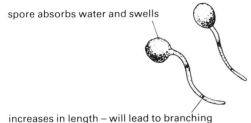

spore absorbs water and swells

increases in length – will lead to branching
and to a radiating network of hyphae

Figure 18   *Mould spore germination*

## Moulds

### Occurrence

Moulds, in contrast to yeasts and bacteria, can often be seen easily with the naked eye. The aerial mycelium often has a height in the region of 500 μm. The typical growth that they display is fluffy and is a familiar sight on damp newspapers, old leather, damp walls, rotting fruits and other foods, such as cheese and jam, and may be black, white or variously pigmented. In further contrast to bacteria and yeasts, moulds are multicellular, being composed of many cells joined together.

Biochemically, moulds are very active, and are primarily saprophytic organisms. Moulds break down complex organic materials into simpler substances and in so doing contribute to the rotting of leaves and other material in the soil. Consequently, in order to prevent rapid deterioration, the bases of fencing posts and wooden telegraph poles must be protected with creosote or other anti-mould compounds.

The same activity contributes to the widespread spoilage of foods, although in some cases mould growth in foods is sought, as when they are used in the ripening of cheeses such as Roquefort and Camembert. A further application of the biochemical activity of moulds is in the ability of some to produce antibiotics – notable among these is the *Penicillium* group of moulds.

A few moulds are pathogenic, causing diseases in plants and in man. The serious respiratory disease 'Farmer's lung' is caused initially by inhaling the spores of the mould *Aspergillus*. Moulds also cause skin infections such as 'athlete's foot' and ringworm.

### Morphology

It is possible to see with the unaided eye that moulds are composed of many threads – the *hyphae*, the mass of which are known as the *mycelium*. Moulds grow by extending the length of the hyphae either at their tips – *apical growth*, or within their length – *intercalary growth*. The hyphae of some moulds have cross walls – *septae*, whose presence is used in preliminary identification of the mould. The hyphae run over and through the medium on which the mould is growing, obtaining nourishment from it, although parts of some hyphae are primarily concerned with reproduction rather than feeding.

### Structure

The cell wall of the majority of moulds is chitinous (complex glucose-amine) but in some it may be composed of cellulose. Definite nuclei are present within the cytoplasm. These nuclei are separated by cross walls in the septate moulds, whereas in the non-septate moulds they are arranged along the length of the hyphae without separation. Within the cytoplasm there may be food substances such as carbohydrates, fats or protein droplets.

### Reproduction

Moulds reproduce by the production of spores by asexual methods or by a mating process (sexual reproduction). In either case a large number of spores is produced by the mould plant. These may be carried away by air currents, animals or water and some survive to produce new individuals (Figure 18) and colonies (Figure 24).

A single hypha may produce many thousands of asexual spores which are more resistant to adverse conditions than the hyphae, but they are more fragile than bacterial spores. Figures 19, 20, 21, 22, 23, 24, 25 and 26 show the formation of asexual spores in several species of moulds.

*Sporangiospores* are produced inside a spore case – the sporangium at the tip of a fertile

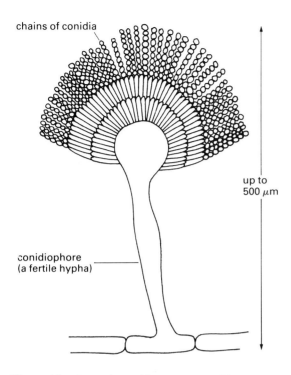

chains of conidia

up to
500 μm

conidiophore
(a fertile hypha)

Figure 19 *Asexual mould spores – conidia, shown here by a member of the genus* Aspergillus. *This group is important because it is able to grow in and on a wide variety of substrates and, as a result, is very destructive. Their mycelium is colourless, pale or brightly coloured. the sporing heads are brightly coloured and 30–50 μm in diameter. The conidiophore is up to 500 μm in length*

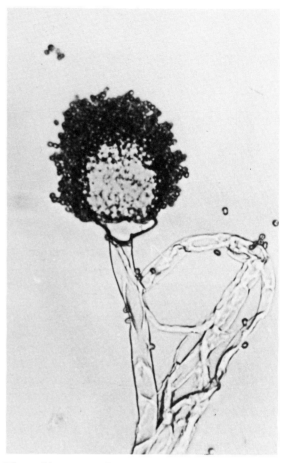

Figure 20 *Aspergillus* (× *1000*)

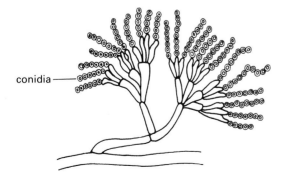

conidia

Figure 21 *Asexual mould spores – conidia. This example of* Penicillium *shows the mode of formation of conidia which are cut-off from the hypha.* Penicillium *species are common, for example the green-blue growth on rotting lemons and oranges, and in cheeses*

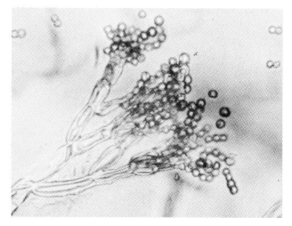

Figure 22 Penicillium (× *1000*)

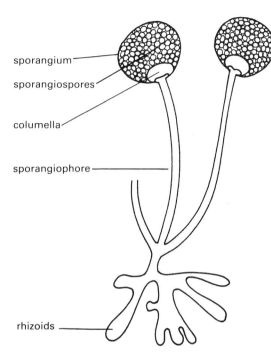

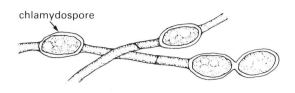

Figure 25 *Asexual mould spores – chlamydospore, a thick walled resting spore formed by a swelling and thickening of single short cells submerged vegetative or aerial hyphae. These are formed in many species, for example* Mucor racemosus

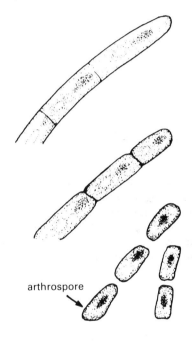

Figure 23 *Asexual mould spores – sporangiospore, shown here in* Rhizopus nigricans. *This is the 'bread mould', which is very common and is involved in the spoilage of many foods*

Figure 26 *Asexual mould spores – arthrospores, formed by fragmentation of the mycelium, for example* Geotrichum candidum

Figure 24 *A large mould colony growing on a nutrient medium*

hypha. The multinucleate protoplasm inside the sporangium is divided by a process of cleavage into many spores which form walls round themselves. *Conidia* are cut off either singly or in chains from the tips of fertile hyphae called *conidiophores* and are not enclosed within a case. *Arthrospores* are formed by hyphae fragmenting into cells. A fourth type, the *chlamydospores*, are formed by the cells of hyphae developing thick walls and changing into spores.

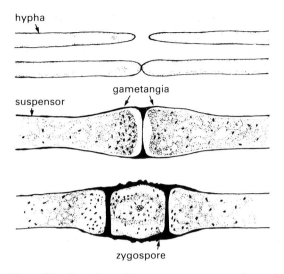

Figure 27   *Sexual mould spores – zygospores, formed by the union of the tips of two hyphae which may come from the same mycelium or from different mycelia. Zygospores are covered by a tough wall which enables them to survive drying for long periods. Species of mucor form zygospores*

Sexual spores are formed in some species of moulds by fusion of two hyphae forming a protective case for the contained spores (Figure 27).

## Classification of micro-organisms important in the food industry

The most common system of classification used by microbiologists today is based on differences in the structure, followed by differences in biochemical ability. Micro-organisms are divided into bacteria and fungi and then within these large groups sub-divided into families, orders, genera and species – more and more precisely defined groups. This system has disadvantages but, nevertheless, it is still widely used. It will eventually be replaced by a

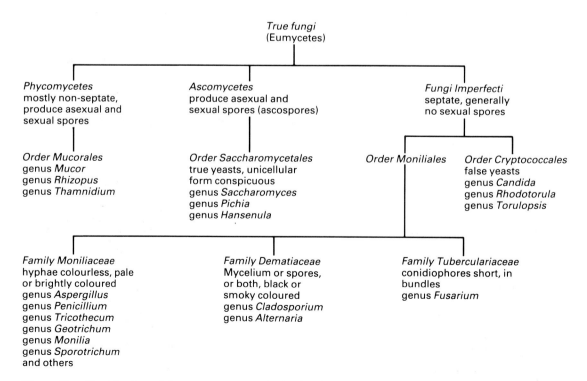

Figure 28   *Classification of fungi important in the food industry*

computer-based classification system known as *numerical taxonomy*. In this system the similarity between organisms is assessed numerically and the organisms are arranged into groups on the basis of the affinities. However, the classification of the organisms shown in Figures 28 and 29 is based on the traditional morphological and biochemical differences.

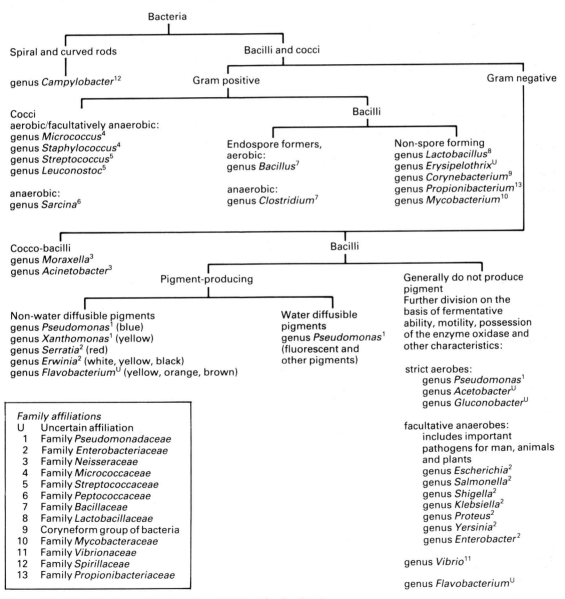

Figure 29 *Classification of bacteria important in the food industry*

# Chapter 2

# Growth and survival of micro-organisms

Because micro-organisms are invisible to the naked eye the term 'growth' tends to be used to mean two things at the same time : cell growth and population increase. These two factors should be thought of as separate but connected events.

## Cell growth

A growing cell is undertaking ordered biochemical activities which result in the production of increased quantities of cellular materials and in the exact replication of the nuclear material (the DNA and RNA). These activities are co-ordinated at a pace which eventually allows that cell to divide and produce two cells.

For growth to occur, a source of suitable materials for the synthesis of the cellular constituents is required, together with sources of energy and sufficient water.

### Materials for growth

All organisms require the elements carbon, hydrogen, oxygen, nitrogen, sulphur, phosphorus, magnesium and iron, particularly, and other elements in trace quantities. Some organisms obtain them solely from inorganic sources, while, more commonly, others obtain them from a mixture of organic and inorganic substances. Large organic molecules from a complex food supply must first be broken down into substances of smaller molecular weight and simpler structure to be taken into the cell across the cell membrane. For this purpose many micro-organisms secrete *enzymes* into the environ-ment. The types and quantities of the enzymes produced dictate the type of organic matter on which micro-organisms grow. *Lipolytic* organisms break down fats and oils, *proteolytic* organisms break down proteins and amino acids, and *saccharolytic* organisms break down sugars and starches. Micro-organisms with these characteristics contribute to the spoilage of foods with high fat, protein or sugar contents respectively. Some organisms are able to break down cellulose – a major constituent of plant cells – and therefore contribute to the decay of both green and non-green vegetables, and to the spoilage of fruits.

The nutrients on which the organisms depend for life pass into the cell by means of several mechanisms. Water molecules, are drawn in by the osmotic pressure of the cell. Some inorganic ions simply pass in down a concentration gradient, but most inorganic ions are selectively and actively carried into the cell by highly specialized mechanisms. Organic materials, if of small enough molecular weight (these are few), may diffuse into the cell, but generally organic materials are carried into the cell actively.

Inside the cell biochemical processes use the nutrients as building blocks (sometimes after chemical changes) from which to synthesize the large molecules of which the cell is constructed. This is known as *metabolism* and it requries energy. Different metabolic activities take place at different sites in the cell, for example protein synthesis from amino acids takes place on ribosomes located on membranes within the cell.

### Sources of energy

Energy for metabolism is derived initially from an external supply, such as sugars, or, in the case of photosynthetic organisms, from sunlight. That energy is then converted into a chemical form inside the cell which can energize the various chemical reactions. In this book we are largely concerned with *heterotrophs* – those organisms which use organic materials as a source of energy. Heterotrophs can grow on a wide variety of food sources which range from the very simple to the very complex.

Thus, whether an organism grows or not is dependent on the substrate providing the nutrients it requires, and whether it is able to break down complex chemicals and transport the nutrients into the cell. The cell must then, of course, be able to metabolize these materials once they are in the cell.

### Water requirement

All life requires water to sustain it. Water is the medium in which the majority of chemical reactions within the cell take place. Water is very important in helping to maintain the spatial relationships of the molecular structure of the cell's components. Water facilitates the inward passage of soluble food stubstances; it is required around the cell to bring food up to it and waste products away from it. Water forms the highest proportion of the cell constituents.

All these activities require water in the *liquid* form, and when water is crystallized in the form of ice, or chemically bound in strong salt or sugar solutions, it is unavailable for use by micro-organisms.

The amount of liquid water available in food or in a solution can be described in terms of water activity – $a_w$. Pure water has an $a_w$ of 1.0. A solution has an $a_w$ of less than 1.0 if solutes (materials in solution) are present. For example, a dilute solution could have an $a_w$ of 0.91, whereas a stronger solution could have an $a_w$ of 0.66. It has been found that different micro-organisms have different requirements for water. On the whole, bacteria can only grow and multiply in a high $a_w$; yeasts survive with less, and moulds with the least. No bacteria, yeasts or moulds can thrive in pure water ($a_w$ of 1.0) because it implies that there is no food present (see Figure 30).

When foods are dried in order to preserve them, use is made of the fact that if micro-organisms are deprived of water they cannot grow. It is important to remember that organisms can survive under dry conditions (see page 123), so that although the food will remain free from spoilage, it will not necessarily be sterile. If, during damp storage conditions, the water activity rises the food will increasingly support the growth of moulds, and later, as the $a_w$ rises further, yeasts and finally bacteria.

Strong solutions of salt and water may exert an osmotic pressure on the microbial cell, withdrawing water and causing dehydration. The physical phenomenon is complex and related to the nature of the cell membrane. Unless a cell is specially resistant to the high osmotic pressures it will lose water, the cyto-

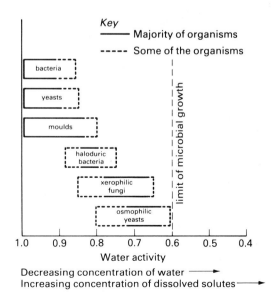

Figure 30 *The effect of water activity on the growth ranges of micro-organisms*
Source: D. A. A. Mossell and M. Ingram, *Journal of Applied Bacteriology*, No. 18, pp. 233–268 (1955); and D. A. A. Mossell, No. 1, pp. 95–118 (1971)

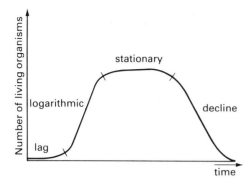

Figure 31 *The phases of growth of micro-organisms*
Notes: 1   The length of the lag phase depends on how 'alien'
the medium is to the organisms, compared with the medium
from which they have come, how favourable the conditions
are for growth and on the metabolic state of the organisms.
2   In the logarithmic phase the numbers of organisms
becomes very large, doubling with every cell division. This
type of increase has the property that when the logarithms of
the numbers are plotted that part of the curve is a straight
line.

plasm will become dehydrated, and the cell will
die. This principle is applied in the preservation
of food by the use of sugars in jams, syrups and
crystallized fruits. A few organisms are specially
adapted to withstand high osmotic pressures and
may in fact prefer to live under such conditions –
these are *osmophilic organisms*, and are primarily responsible for the microbial spoilage of foods
with high sugar content.

Some bacteria are salt tolerant – *haloduric*,
while others require high salt conditions – these
are *halophilic organisms*. These organisms can
give rise to the spoilage of salted or brined
foods.

## Population change

Population increase in micro-organisms depends
on individual cells undergoing the processes of
reproduction at a faster rate than cells are dying.
Bacteria reproduce by virtue of each cell
dividing into two identical parts – each being a
new individual. This is known as *binary fission*
and involves the replication (doubling) of the
nuclear material followed by the inward growth

of the cell wall to divide the cell into two equal
parts (see page 21). One parent cell has
produced two daughter cells – hence the
population has doubled. It is an asexual method
so a population of thousands can arise from a
single original organism. The rate at which cell
division happens is dependent on the environmental conditions in which the micro-organisms
live. When optimum conditions for growth are
provided, increase occurs at the maximum rate.
For some bacteria, for example *Clostridium
perfringens*, the time lapse between one cell
division and the next can sometimes be as little
as 10–12 minutes. However, that rate of cell
division cannot be sustained for a prolonged
period because the environment surrounding
each cell changes. Often the source of nutrients
is used up and the waste products produced by
the many cells may gradually poison the
environment, with the result that the population
undergoes a series of changes which can be
described as the 'phases of growth' (see Figure
31).

### Phases of growth
*Lag phase*
If a cell-free medium is inoculated with cells for
a time varying from a few minutes to several
hours, there may be no cell multiplication while
the cells become accustomed to their environment. During the latter stages of this phase each
cell increases in size.

*Logarithmic phase*
When the cells are adapted to the medium, they
multiply at regular intervals until the maximum
number that can be supported by the environment is reached (see Figure 32).

*Stationary phase*
Growth slows down during this phase because of
the depletion of food, accumulation of waste
products, overcrowding, and other factors. Then
individually the cells begin to age and the
number of cells dying gradually increases until it
equals those produced, resulting in no overall
increase in the living cell numbers.

| Time interval between each cell division | Total time passed | Total number of cells present (a) | Number of cell divisions | log₁₀ (b) | log₂ (c) |
|---|---|---|---|---|---|
| 10 minutes | 0 minutes | 1 | 0 | 0.00 | 0 |
| | 10 minutes | 2 | 1 | 0.30 | 1 |
| | 20 minutes | 4 | 2 | 0.60 | 2 |
| | 30 minutes | 8 | 3 | 0.90 | 3 |
| | 40 minutes | 16 | 4 | 1.20 | 4 |
| | 50 minutes | 32 | 5 | 1.50 | 5 |
| | 60 minutes | 64 | 6 | 1.80 | 6 |
| | 70 minutes | 128 | 7 | 2.10 | 7 |
| | 80 minutes | 256 | 8 | 2.40 | 8 |

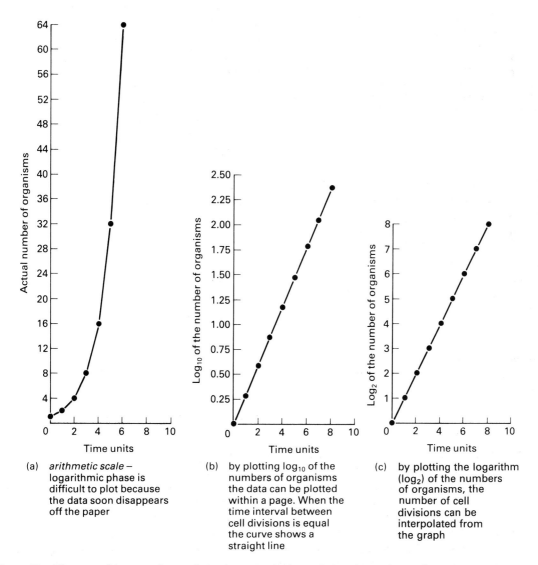

(a) *arithmetic scale –* logarithmic phase is difficult to plot because the data soon disappears off the paper

(b) by plotting $\log_{10}$ of the numbers of organisms the data can be plotted within a page. When the time interval between cell divisions is equal the curve shows a straight line

(c) by plotting the logarithm ($\log_2$) of the numbers of organisms, the number of cell divisions can be interpolated from the graph

Figure 32   *The type of increase in numbers shown in the logarithmic phase of growth*

*Phase of decline*

If cells are not transferred to a new favourable environment they will gradually die, leading to the death of the whole culture. The surroundings which were once ideal have been changed by the organisms to such an extent that none can survive.

The time period over which these changes take place varies considerably. The cycle can be as short as 48–72 hours, or at the other extreme, prolonged over several weeks.

*Conditions influencing the growth rate*

A species of micro-organism will grow quickly or slowly to high or low maximum numbers dependent on the physical and chemical conditions. The growth curves of the same organism under two sets of conditions can be very different. A species can only grow if certain concentrations of a minimum range of nutrients are present, from which it can synthesize all the other chemicals it needs for growth and reproduction. If a wider range of suitable nutrients is provided, perhaps also in greater concentration, growth is usually faster (see Figure 33).

Variation in other environmental conditions has a comparable effect. Growth occurs faster and to higher maximum numbers under some conditions than under others.

*The effect of temperature on growth*

It can easily be demonstrated in the laboratory that there is a *minimum* temperature for the growth of a particular organism. Temperatures above the minimum stimulate growth and cause more rapid cell division. This occurs up to the

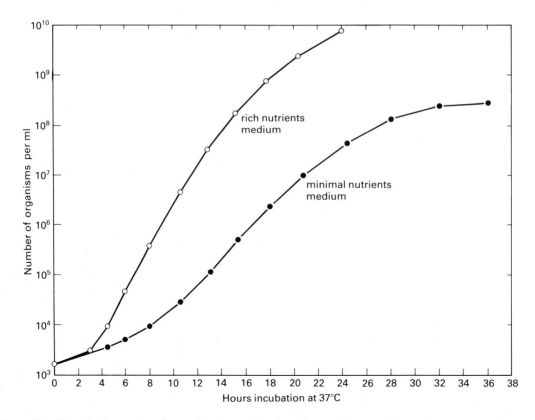

Figure 33 *Growth of a species of organism in a minimal nutrient medium, and in a rich nutrient medium at the same incubation temperature*

Table 2

|  | Optimum temperature for growth | Approximate growth range |
|---|---|---|
| Psychrophiles (cryophiles) | + 10 to + 15°C | − 3 to + 25°C |
| Mesophiles – human pathogens | + 35 to + 42°C | + 6 to + 45°C |
|         – saprophytes | + 18 to + 23°C | + 6 to + 45°C |
| Thermophiles | + 45 to + 55°C | + 30 to + 65°C |

*optimum* temperature – the temperature at which the organism grows best and fastest. Beyond that, growth is adversely affected and the high temperatures discourage growth to a point where a *maximum* temperature for growth can be defined. Above the maximum and below the minimum temperature multiplication does not occur, neither in the laboratory media nor in the natural environment outside the laboratory (see Figure 34).

A classification of organisms based on their temperature requirements and preferences is listed in Table 2.

*Psychrophiles* occur naturally in cold places such as the Arctic, the North Sea and the Atlantic Ocean. They are not only present in sea water but also on and within fish, and in soil. Such organisms are important in refrigerated and frozen storage where they may grow, multiply and possibly spoil the stored food. Examples are strains of the moulds *Penicillium*, *Cladosporium* and *Neurospora*, bacteria of the genera *Pseudomonas*, *Flavobacterium*, *Bacillus* and *Clostridium*.

*Mesophiles* form a large group including those which live on and in the warm bodies of warm blooded animals. Some mesophiles are pathogenic (cause diseases) to man and animals, and many of these have an optimum temperature for growth of 37°C, which corresponds with the human body temperature. Other mesophiles live in the soil and in inland water in temperate climates – for example the saprophytes which use decaying organic matter as a source of food.

*Thermophiles* prefer higher temperatures than

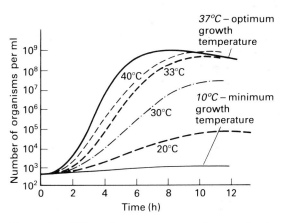

Figure 34 *Effect of temperature on the growth of a species of mesophilic bacterium*

mesophiles for growth and are found where the ambient temperature is within their growth range.

### The effect of pH on growth

pH is a scientific term describing in numbers the acidity or alkalinity of a fluid. Micro-organisms can only grow and multiply within a certain pH range. The vast majority of organisms, the *neutrophiles*, prefer to live in a neutral environment near pH 7. *Acidophiles* include a small group of organisms preferring an acid medium and which do not thrive as well in the neutrophilic range. When food is preserved by pickling, the acid environment provided by the vinegar protects the food from spoilage by neutrophiles, although spoilage by unwanted acidophiles can still occur (see Figure 35).

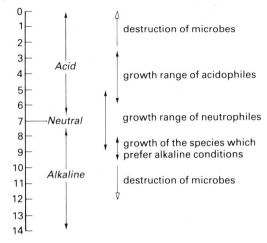

pH scale

Figure 35 *The effect of pH on the growth range of micro-organisms*

### Availability of oxygen

All organisms respire – that is, they obtain energy by breaking down certain chemicals, usually sugars, inside the cell. Some organisms use a process which requires oxygen obtained from the environment. These are called *aerobic organisms*. Some can only release energy in the absence of oxygen. These are called *anaerobic* organisms. A third kind of organisms are adaptable and release energy for their own use either in the presence or absence of oxygen. These are called *facultative organisms*.

Specific gaseous environments influence the type of flora which is supported by an environment. Where there is an abundance of oxygen aerobes will thrive; where oxygen is lacking the facultative and anaerobic organisms will grow and multiply.

To summarize, the rate at which micro-organisms reproduce, or whether they grow at all, depends on the physical and chemical qualities of the medium.

### Products of growth

During its lifetime, due to its growth activities, a micro-organism produces waste products which are either retained inside the cell and released on cell death or, more importantly, secreted into the medium in which the organism is living.

Examples of these are:

Gases – carbon dioxide, hydrogen sulphide
Acids – lactic, propionic, acetic
Alcohols
Aromatic compounds – flavours, off-flavours, taints
Antibiotics
Toxins

This process has traditionally been exploited by allowing certain micro-organisms to grow and help in the production of foods, such as wine, beer, sauerkraut, vinegar, yoghurt, fermented meats, cheeses, pickles, baker's yeast and a variety of Asian sauces and spices. In other circumstances microbial growth and by-products cause the symptoms of spoilage in food stuffs.

Today, the chemical abilities of micro-organisms are being further exploited in the 'new' science of *biotechnology*. New processes which involve the growth of micro-organisms on an industrial scale are being used to produce a wide range of food additives and processing aids. Examples of these are:

The production of citric acid
Monosodium glutamate (MSG)
Flavour enhancers – the nucleotides, 5' inosinic acid and 5' guanilic acid
Amino acids, such as lysine
Starch derivatives including isoglucose and xantham gum
Enzymes, such as microbial rennets and ox-idoreductase
Oils and fats

Although micro-organisms produce many useful by-products, they can be further exploited by inducing them to grow on the waste products of industry – petroleum, agricultural and chemical wastes. Excesses, such as starch wastes (flour, grain, potato water), lactose in whey and cellulosic waste from the paper-processing industry, can be the substrates on which various micro-organisms grow. The mass of micro-organisms which results (the *biomass*) is then a basic material from which protein can be extracted to supplement human or animal foods.

This protein is sometimes called *single cell protein* (SCP); or *mycoprotein* if it comes from mould or yeast growth.

Certain proteins, peptides and amino acids are already being produced from microbial sources (bacteria, yeasts and moulds). Several big companies, such as Ranks Hovis Mac-Dougall, Nestlé and ICI, have invested millions of pounds in the research and development work necessary to develop the basic idea of the production of microbial protein.

Hydrolysed proteins derived from bacterial or yeast protein are widely used as flavour enhancers. Amino acids can be formed from hydrolysed protein, or by microbial fermentation processes. For example, glutamic acid and lysine are produced in a fermentation process using *Corynebacterium glutamicum*.

Continuing research will allow the improvement of all these processes by using, finding or making micro-organisms which do the particular jobs better than the previously used organisms. It is already possible to transfer genes, or groups of genes (gene sequences), from one species of microbe to another with the subsequent expression of that gene in the host species. This development of new strains is achieved by recombinant DNA techniques – genetic engineering. In the food industry this will probably mean that expensive food additive materials, such as amino acids and vitamins, will be manufactured more cheaply than is currently possible, by growing micro-organisms and extracting the products from them.

Many antibiotics, which are the by-products of growth, have been produced for years on an industrial scale. The process involves the large scale growth of moulds, followed by the careful extraction and purification of the antibiotics.

Finally an unwanted expression of growth is the production of toxins in foods. The toxins are usually produced maximally during active growth. Where growth conditions are not very permissive, the production of toxin tends to be inhibited, although for safety's sake it is better to consider that any condition which permits growth of a toxigenic organism also permits toxin production and accumulation.

## Microbial survival

Vegetative cells which find themselves in an environment which does not support their growth can often survive for quite a long period. In a population of cells, individuals die at different rates with the result that the overall number of living organisms decreases gradually.

A growing organism will direct its metabolism towards eventual cell division. A surviving organism 'ticks-over' at a rate which does not permit metabolic activity to be directed towards cell division. If a surviving organism is transferred to a better environment it can be revived and undertake growth again. Vegetative cells as well as spores can survive for periods of time which can be extensive – months, or even years in some circumstances. The rate at which death occurs in the surviving population is, as growth is, dependent on a number of environmental factors – the specific combinations of which either accelerate or decelerate the death rate. The factors are the same as those which affect growth, especially the availability and the types of nutrients, temperatures, pH, $a_w$ and oxygen tension.

The majority of organisms grow in an aqueous environment ($a_w$ of 0.99–0.90) containing dissolved nutrients and gases. Within the cells a slightly lower $a_w$ than the external one is maintained, and the osmotic pressure is such that they remain fully hydrated. However, in *very pure water* nutrients will be in very short supply and the cell will tend to be over-hydrated. In these conditions the cells may burst when first introduced into the water; surviving that, they starve gradually. The organisms in sewage suffer this fate when sewage effluent is emptied into rivers and seas. Delicate pathogens–salmonellae, shigellae, *E. coli* and others – tend to die out rapidly (in a matter of hours); saprophytic soil organisms survive longer. The concentration of suspended and dissolved organic matter is very influential in these circumstances – high concentrations tend to protect and prolong or even promote microbial viability.

Conversely, in conditions where solutes are concentrated, cells begin to suffer from the risk of excessive water loss and toxic effects of specific solutes present. Below the $a_w$ tolerated for growth, even if other suitable nutrients are present, the cell population diminishes.

In those *intermediate water foods* (jam, dried and salted meats and fish, cakes, semi-moist pet foods, dried figs etc.) whose $a_w$ value lies in the range of 0.70–0.90, there is a tendency for surviving mould and yeast spores to spoil the products by growth if the moisture content accidentally rises. However, bacterial pathogens, such as salmonellae, can also survive in intermediate water foods, and are potentially capable of causing salmonellosis (see Chapters 4 and 5) on the rehydration and consumption of the food. There is evidence that salmonellae can persist for long periods in intermediate water foods – for example, for over a year at 5°C in concentrated solutions of sucrose (up to 66% weight for weight (w/w)); in 20% w/w sodium chloride in nutrient broth for over 70 days at 5°C, and up to 30 days at 20°C.

*Dried foods* kept dry keep for long periods because they do not allow the growth of micro-organisms. However, the drying process itself does not kill many organisms but rather tends to concentrate them. The resulting dried product can contain many living cells which survive for long periods. In dried milk powder, for example, not only do spores survive, but also vegetative cells, such as those of streptococci. Extreme dryness allows for prolonged survival. When combined with freezing (freeze-drying) it provides a method for almost permanent storage, as is applied in the preservation of stock cultures of micro-organisms.

*Cold storage* tends to preserve micro-organisms, but where they occur in fresh foods which are subsequently frozen, the manner of freezing and the rate of cooling very substantially affect both the number that are killed initially, and the rate at which the remaining cells die. In general terms, raw frozen foods are never sterile. The thawing rate also affects cell survival. At storage temperatures below −70°C, little to no microbial death occurs. Between −60 and 0°C, the percentage survival of most species decreases with time, the death rate depending

on the storage temperature, the freezing menstruum and the number of cells present.

Spores, in contrast to vegetative cells, have an inherent capacity to survive in most environments which kill the vegetative cells which produce them. Spores can remain dormant for long periods – this is true for bacterial, yeast and mould spores. In dry conditions spores may survive for years, although the level of viability is very much dependent on the precise environmental conditions. In general, bacterial spores are capable of survival over longer periods than mould and yeast spores; and in all cases a protracted but definite loss of viability occurs.

### Injury and repair in surviving cells

As indicated, microbial cells can withstand a variety of adverse storage conditions. They also withstand 'moderate' exposure to heat, and ultra-violet and ionizing radiations. However, they do not survive unscathed – they suffer injury. The injury is shown when the cells are eventually taken out of their storage condition and transferred to an environment in which it would be supposed that they could grow. It is not easy to generalize about the reactions then shown. In careful, controlled laboratory conditions, injured cells can be recovered from storage in a frozen, dried or heat-injured condition, and the injury they have suffered can be demonstrated as an inability to grow straightaway – a prolonged lag period occurs. During this period the injury or injuries suffered are repaired before cell multiplication is possible. Subsequently, cell growth, cell division and normal population increase occur. Injury can also sometimes be shown by the cells' inability to grow in the normally used selective media.

In practical terms there are two important aspects: firstly, stored foods of the sorts mentioned earlier must be assumed to contain living organisms; and secondly, microbiological techniques for quality assessment must take the injury of the stored cells into account. If injury is ignored, quality control counts can indicate low counts when in fact the test food contains a high number of injured cells which simply have not been shown by the technique used.

# Chapter 3

# Destruction of micro-organisms

The most important ways of killing microbes are by the use of *heat* and *chemicals*. Among the other ways of killing microbes are *irradiation*, *ultrasonic sound* and *very high pressure*, none of which has great significance yet in the food industry although irradiation may play an increasing role.

Heat, chemicals and irradiation destroy cells in different ways and are used in different circumstances. Each of these will be considered separately, although there are factors in common with all of them. These are:

The number and varieties of organisms present.
How the living population responds to the killing agent.
The time of exposure.
The relation between the temperature and the concentration of agent used.
The environmental conditions of exposure; the pH, presence of organic material, density of menstruum.

Cells are destroyed by the coagulation of proteins, especially enzymes (wet heat); by severe loss of water and burning of cells (dry heat); by disruption of the transport of substances into and out of the cell, by metabolic disturbances and by oxidative reactions (chemicals); and by molecular reorganizations, the degree of which determines whether the molecules can continue to function normally and hence whether the cell survives (irradiations).

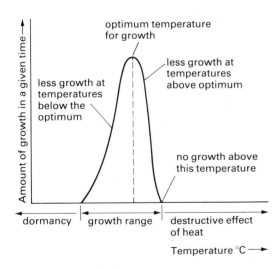

Figure 36 *The effect of temperature on the viability and growth of an organism*
Note: Beyond the maximum growth temperature, increases in temperature have a destructive effect, but at temperatures below the minimum growth temperature the culture becomes dormant and some of the population will probably survive until the temperature and other conditions are suitable for growth again.

## Heat

*Mode of action*
As seen in the previous chapter, temperature affects the rate at which micro-organisms grow, divide and increase in number. Excesses of heat beyond the growth range have adverse effects on cells which lead to their death (see Figure 36). When heat is applied in the presence of water, and above the maximum temperature for growth, cytoplasmic structures such as ribo-

somes and membranes are damaged, structural and enzymic molecules are denatured, first of all reversibly, and then irreversibly, and metabolism cannot proceed.

Heat applied in the presence of water is known as *wet heat*. If heat is applied in the absence of water the cells are destroyed by their dehydration and oxidation. Such heat is known as *dry heat*. Sterilization is defined as any process (by heat or other method) which kills or removes all microbial life. Sterilization by dry heat generally requires a longer time than by wet heat to achieve the same results.

### Effect in cell populations
When a culture of cells is exposed to heat the cells do not all die at once, so the greater the initial number, the longer it takes to sterilize the culture. This is partly due to the location of the cells and the time it takes for the heat to penetrate the suspending medium at killing level, and partly due to the differences in susceptibility of the cells. The term 'heat sensitive' is used to describe a culture of cells in which the majority (perhaps 99.9 per cent) are destroyed by exposure to wet heat at 63°C for 30 minutes, or by conditions less severe than this.

The term 'heat resistant' is usually applied to a culture of cells in which the majority can only be destroyed by wet heat at 100°C for 10 minutes, or by conditions more severe.

Between the two extremes, strains of organisms may be described as being *thermoduric*; in order to destroy the majority of cells, conditions more severe than 63°C for 30 minutes, but less severe than 100°C for 10 minutes are required (see Figure 37).

The term heat resistant is best applied to bacterial spores – a form of organism which can survive a variety of adverse circumstances, being provided with the special protection of a spore coat. In order to destroy spores, heat in the region of 100°C and usually above this in moist conditions is required. Individual spore types vary in their heat resistance – some being sensitive to exposure at 100°C for 10 minutes; others requiring the extreme of exposure at 121°C for several hours; while the majority of

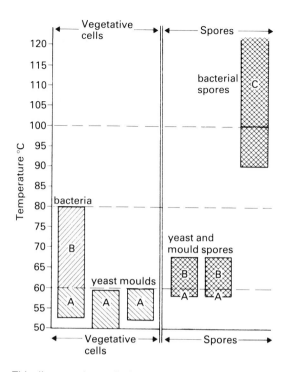

This diagram shows the broad categories of heat sensitivity of the micro-organisms

Group A    heat sensitive organisms – examples are members of genus *Salmonella*, genus *Escherichia*, genus *Proteus*, genus *Pseudomonas*, vegetative cells of genus *Bacillus* and genus *Clostridium*; vegetative yeast cells, most mould mycelia, yeast spores and some mould spores.

Group B    thermoduric organisms – examples are some members of genus *Micrococcus* associated with milk spoilage, *Streptococcus thermophilus* and other streptococci, members of genus *Lactobacillus*, most asexual mould spores, and sexual yeast and mould spores, most viruses.

Group C    very heat resistant organisms – the primary examples are the spores of genus *Bacillus* and genus *Clostridium*

Figure 37    *Heat sensitivities of micro-organisms*

species are destroyed by exposure at 121°C for 15 minutes. (see Figures 37 and 38).

In the food industry certain spore types are of particular significance. If they are given the opportunity to germinate and the vegetative cells produced multiply, they either cause illness in a consumer or spoilage of the food. Sterilizing processes are designed to destroy them and at the same time destroy other less heat resistant organisms. Of particular importance are:

1  Spores of *Clostridium botulinum* – the toxin produced by the vegetative cells causes a very severe form of food poisoning.
2  Spores of *Bacillus stearothermophilus* – the vegetative cells of this organism can cause a type of spoilage known as 'flat souring' in low acid canned foods.
3  Spores of *Clostridium perfringens* – if the vegetative cells occur in food under the right conditions they can multiply and cause food poisoning. There are two groups of strains. The spores of both groups survive normal cooking processes, but they differ in their degree of heat resistance. Adequate destruction of the spores of the heat sensitive strain requires exposure to moist heat at 100°C for between 30 and 60 minutes; the heat resistant spores can withstand moist heat at 100°C for over 60 and up to 240 minutes.

### Number of organisms

The higher the number of living cells present, the longer the time it takes to destroy them all. Differences in heat sensitivity among individuals and between strains, and the greater heat resistance of spores mean that where the number and variety of contaminating organisms are high, more severe heating processes are required to achieve sterility than when the numbers of organisms are low and the types limited.

### Time/temperature effects

Irrespective of whether wet or dry heat is used as a method of disinfection or sterilization, the following considerations influence how the heat is applied.

When micro-organisms are heated, the number of minutes and the temperature of exposure required to destroy them are linked inversely – the longer the exposure time the lower the temperature required, provided of course that the temperature used is above the growth range. Conversely, the shorter the time, the higher the temperatures required (see Figure 39). Under

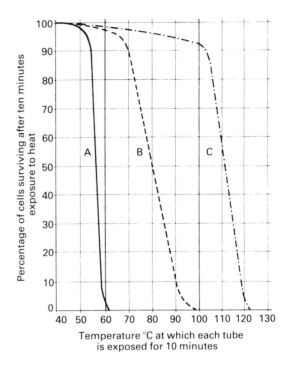

A number of tubes containing each of the cultures A, B and C are used. The members of each set of tubes are held for 10 minutes at different temperatures in the range shown – 40°C to 130°C – after which each is examined to ascertain the surviving viable numbers. From the numbers in control samples taken prior to heating, and from surviving numbers, it is possible to calculate the percentage of cells which have survived the heat treatment. The results are plotted as a graph. Thus A is shown to be *heat sensitive*, B is more resistant to heat – *thermoduric*, and C the most *heat resistant*

Figure 38  *To show the effect of heat on three different types of organism*

dry conditions longer times and higher temperatures are required to achieve the same degree of kill as with wet heat.

Dry heat is only practicable as a sterilizing agent where the item being treated will not suffer physical damage through the heating process, and is generally used in the laboratory for sterilization of glassware or other similar material; whereas wet heat is used very widely in the laboratory, in kitchens and in food factories.

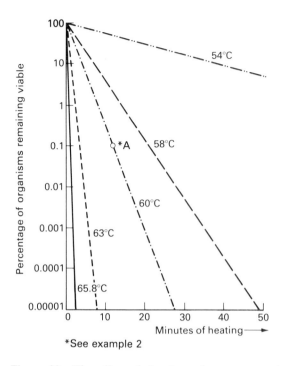

*See example 2

**Figure 39** *The effect of the time of exposure and temperature of exposure on the viability of a heat sensitive organism, for example salmonella in an aqueous system*
Note: The application of high temperatures destroys the organisms in a short time; the use of lower temperature requires a longer time to achieve the same level of destruction.

### Decimal reduction time

Up to now a vague term –'the same degree or level of kill' – has been used in the text but there are more precise terms to describe how effective heat is as a killing agent. On exposure to lethal heat, a population of microbial cells dies out logarithmically, that is a constant temperature of exposure will reduce the viable population by a constant factor in a constant time period.

### Example 1

Suppose $10^6$ organisms/gm are reduced to $10^5$ organisms/gm in 7 minutes at 63°C, then, the living population of $10^5$/gm will be diminished to $10^4$/gm after a further 7 minutes at 63°C. Again the $10^4$ organisms/gm will be reduced to $10^3$

organisms/gm after 7 more minutes at 63°C. The necessary time period to achieve each ten-fold reduction in the population is known as the *decimal reduction time* (D). In this example the decimal reduction time at 63°C ($D_{63°C}$) = 7 minutes.

### Example 2

*Staphylococcus aureus* is one of the less heat sensitive species of non-sporing organisms. But its sensitivity depends on what it is heated in – in this way it is like other organisms. At 60°C in phosphate buffer, an exposure of 0.5 to 3.0 minutes is required to reduce the population to one-tenth ($D_{60°C}$ = 0.5 to 3.0 minutes). But in milk containing 57 per cent weight per volume (w/v) sucrose, to reduce the population to one-tenth, at 60°C, 42 minutes is required ($D_{60°C}$ = 42 minutes). Thus if a food contained $10^5$ cells of *Staphylococcus aureus* per gram, and it was desired to reduce the population to less than $10^2$ cells per gram a 3D process would be required.

$$10^5 \rightarrow 10^4 \rightarrow 10^3 \rightarrow 10^2/g$$
$$\quad 1D \quad\quad 1D \quad\quad 1D \quad\quad\quad\quad \text{total 3D}$$

In phosphate buffer this would require a maximum of 9 minutes (3 × 3) at 60°C; in milk plus 57 per cent w/v sucrose it would require 3 × 42 minutes (126 minutes) at 60°C. For another organism, of different heat sensitivity, the time required at 60°C to produce a 3D kill would be different – in Figure 39 the data indicate that 12 minutes at 60°C would reduce the population to 0.1 per cent of its original value (point A – 3D kill) in that aqueous system.

Table 3

| Initial number | Dose of treatment | Number of cells left |
|---|---|---|
| $10^7$/gm | 1D | $10^6$/gm |
| $10^7$/gm | 7D | 1–9 cells/gm |
| | | = 1–9 × $10^2$ cells/100 gm |

So, the effectiveness of heating processes can be more precisely described by the number of decimal reductions achieved.

The reduction of the numbers of the viable cells (or spores) by a constant fraction always leaves a small viable number. An example is shown in Table 3.

If a further 2D treatment is given, the population would be reduced to 1–9 cells/100 gm or an average of 0.01–0.09 cells/gm. The latter is not possible since cells are discrete bodies. At worst 9 separate 1 gm quantities would contain 1 cell each, and 91 would be free of living organisms. Packaged separately, 91/100 would be sterile. Packaged in one 100 gm unit, the unit would not be sterile.

So, in practice, small quantities can more assuredly be freed of viable cells than large ones because cells are discrete, and the one or two remaining at the 'end' of a process will be contained within separate small volumes, leaving the other small volumes free.

Fortunately, exposure at progressively higher temperatures dramatically reduces the time required to achieve the same number of decimal reductions (see Figure 39). If the D-values at different temperatures are plotted the same dramatic effect as the temperature increases can be seen more clearly (see Figure 40). The same principles apply to the destruction of bacterial spores but the temperatures involved are close to and generally above 100°C.

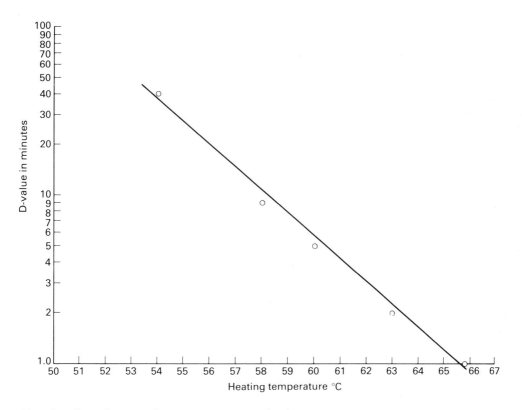

Figure 40   *The effect of increased temperature on D-values**
*Note*: Data from Figure 39 are replotted to show that small increases in temperature dramatically shorten the time required to reduce the population to 10 per cent of its original value, the *D-value.

## Chemicals

Substances which have an adverse effect on micro-organisms, preventing cell growth or in some way leading to cell death, are known as anti-microbial agents (AMA). These include disinfectants, sanitizers, chemotherapeutic substances and antibiotics. These are all defined in Table 4.

Some AMA occur naturally in microbial habitats where their effect is to limit the number and range of organisms which are able to grow and thrive. Naturally occurring and man-made agents are used in medicine to overcome infections; others are used in the hospital environment to check the spread of infection, and in food factories and other food premises to prevent the spread of food poisoning and food spoilage organisms. Table 5 contains a list of a wide range of chemically unrelated compounds and some of the uses to which their anti-microbial action is put.

### *Mode of action*

Lethal effects of AMA depend on their coming into contact with the cell wall and cell membrane, both of whose physical structure may be affected in such a way that their specific functions can no longer be performed. For example, cell constituents may leak out, or external substances which may normally be excluded may be able to enter the cell and disturb the cell metabolism.

The AMA, once inside the cell, may act adversely in one of at least three ways:

1  By interference with enzyme activity, resulting in the disruption of vital processes such as respiration, or synthetic activity.
2  By 'blockage' of vital metabolic reactions

Table 4  *Definitions of some anti-microbial agents*

| Agent | Definition |
| --- | --- |
| Antibiotic | A substance produced in the normal growth of some organisms and which suppresses the growth of others; some are used in chemotherapy. |
| Antiseptic | A chemical used in antisepsis – that is, the destruction of micro-organisms, but not bacterial spores, on living tissues; not necessarily killing all micro-organisms but reducing them to a level not normally harmful to health. |
| Chemotherapeutic substance | A substance used in medicine to destroy pathogenic organisms inside the body. |
| Disinfectant | A chemical used in disinfection (see page 147). |
| Disinfection | The chemical destruction of micro-organisms, but not usually bacterial spores by, for example, phenolic compounds; not necessarily killing them all but reducing them to acceptable levels. The term is applicable in non-medical contexts to the treatment of inanimate objects but today is very rarely used in relation to food contact surfaces. It continues to be used in relation to the treatment of the skin and body membranes. |
| Sanitizer | A chemical used in sanitizing or disinfection. |
| Sanitizing | A process, chemical and other, applied both with, and without heat, which cleanses plant, etc. resulting in microbial counts that are at acceptable levels. It is a more comprehensive term than disinfection. |

See also Chapter 10.

preventing the synthesis of chemicals essential to normal growth.

3  By interference with the genetic mechanisms – the DNA (deoxyribose nucleic acid) – leading to disrupted growth or the inability to reproduce and, in turn, to cell death.

Table 5  *Examples of anti-microbial compounds*

| Group | Examples | Possible uses/comments |
|---|---|---|
| Alkalis | Caustic soda | CIP |
| Acids | Strong inorganic acids | Highly bactericidal. Difficult to handle safely. Removal of milk stone |
|  | Organic acids | Food preservation – natural (lactic acid) or added (propionic) |
| Salts | Sodium chloride | Food preservation |
|  | Sodium nitrate/nitrite |  |
| Metal salts in solution | Copper sulphate | Antifungal – used in agriculture |
|  | Silver nitrate | Mild antiseptic |
|  | Merthiolate | Antiseptic |
| Chlorine compounds | Hypochlorite | Sanitizer in the food industry |
|  | Chlorine gas | Disinfection of water supplies |
|  | Dichloroisocyanuric acid | Swimming pool disinfection |
| Iodine compounds | Potassium iodide | Antiseptic |
|  | Iodophors | Sanitizers in the food industry |
| Alcohols | Ethyl alcohol | Antiseptic. Bactericidal action increases with increased molecule length up to about 8–10 carbon atoms |
|  | Isopropyl alcohol |  |
| Aldehydes | Formaldehyde gas | Fumigant |
|  | Glutaraldehyde | Aldehydes effective against vegetative cells and spores |
| Ethylene oxide | Ethylene oxide gas | Care in use needed. Use in sterilization of heat sensitive plastics. Some residual toxicity remains after treatment |
| Phenols | Phenols | Non-food use disinfection |
|  | Cresols | Strongly smelling |
|  | Coal tar derivatives |  |
| Surface active agents | Ampholytic agents e.g. alkylated amino acids | Surface active sanitizers used in cleaning food plant |
| Cationic detergents | Quaternary ammonium compounds (QAC) | Food plant sanitizing |
| Antibiotics | Penicillin | Medical chemotherapy |
|  | Streptomycin |  |
|  | Novobiocin |  |
|  | Ampicillin |  |

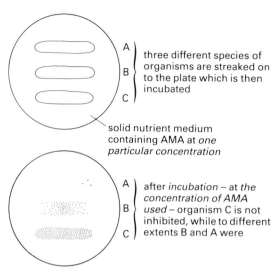

three different species of organisms are streaked on to the plate which is then incubated

solid nutrient medium containing AMA at *one particular concentration*

after *incubation – at the concentration of AMA used –* organism C is not inhibited, while to different extents B and A were

The AMAs effect against a large number of species could be screened by using multipoint inoculation. In any experiment suitable controls have also to be set up to ensure that the effects observed can be interpreted properly. In this experiment the control would be a duplicate plate set up in the same way, but not containing any AMA. This would confirm the ability of each species to grow on the medium and that inhibition of growth was solely due to the AMA

Figure 41   *Testing the anti-microbial spectrum of an AMA – the range of species of organisms against which the AMA is effective*

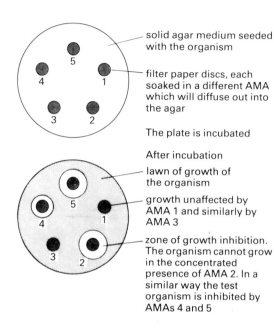

solid agar medium seeded with the organism

filter paper discs, each soaked in a different AMA which will diffuse out into the agar

The plate is incubated

After incubation

lawn of growth of the organism

growth unaffected by AMA 1 and similarly by AMA 3

zone of growth inhibition. The organism cannot grow in the concentrated presence of AMA 2. In a similar way the test organism is inhibited by AMAs 4 and 5

Figure 42   *To test the sensitivity of an organism to a range of AMAs*

Although capsules may offer some protection, vegetative cells do not usually possess any particular structure which gives protection against noxious substances in the environment. Rather, resistance may be due to the inability of the chemicals to enter the cells, or to their local destruction by enzymes secreted by the cells in the surrounding medium. However, it is noticeable that gram positive and gram negative cells often differ markedly in their sensitivities – certain AMA only being effective against gram positives, others only against gram negatives. This is believed to be due to structural differences in their cell walls and membranes.

Bacterial spores are very difficult to destroy by chemicals because they possess thick protective spore coats. There are only a few effective sporicidal agents which can be used without at the same time producing adverse effects on the contaminated surface. For example, concentrated sulphuric acid is sporicidal, but because of its corrosiveness its uses as a sporicide are limited.

The *anti-microbial spectrum* of an agent refers to the range of organisms against which it is effective, no agent being effective against all types of organisms. Sanitization can therefore only be achieved when an agent to which the target organisms are sensitive has been used. Figure 41 illustrates how an AMA may not be effective against all the organisms with which it is confronted. Equally each species of organism is sensitive or resistant to a unique range of AMAs (see Figure 42). The tests shown in Figures 41 and 42 are very simple. They are quite a useful approach for screening purposes – that is finding out whether the problem organisms might be sensitive, or which AMA might potentially be best to use. But the tests, as they stand, do not help very much in working out the practical application of the AMA. Other techniques have to be used for that purpose.

Table 6

| Agents | Generic term |
| --- | --- |
| *Inhibitory to* | |
| Bacteria | Bacterio*static* compounds |
| Moulds | Myco*static* compounds |
| Fungi | Fungi*static* compounds |
| *Lethal to* | |
| Bacterial cells | Bacteri*cidal* compounds |
| Fungal cells and spores | Fungi*cidal* compounds |
| Viruses | Viru*cidal* compounds |
| 'Germs' (disease causing) | Germi*cidal* compounds |
| Bacterial spores | Spori*cidal* compounds |

### Effect of AMA in cell populations

When a strain of organisms is described as being *resistant* or *sensitive* to an AMA, this description presupposes that certain conditions regarding the concentration and the time and temperature of exposure to an AMA are fulfilled.

Anti-microbial agents are either *inhibitory* or *lethal*. Inhibitory agents are those which stop growth and multiplication rather than totally destroy the cells. Should the substance be removed by dilution or inactivation before the cells have aged, the cells present will be able to grow again. Lethal agents kill cells, and may do so in a matter of minutes. (See Table 6.)

When an AMA is used the cells which are *more resistant* than the majority in the conditions under which the AMA is applied are the more important. This is because the majority reaction of a strain of organism which is normally sensitive to an AMA can be changed by the purposeful or accidental selection of the more resistant cells in the population, and by the subsequent provision of conditions under which they can grow and multiply. For example, suppose an AMA at a certain concentration kills 99.9 per cent of the cells and leaves behind 0.1 per cent unaffected. Provided that this 0.1 per cent can find nutrients on which to feed they will grow and increase in number (see, for example, Figure 41 – organism A). The majority will be resistant to the AMA in the concentration in which it was previously used – so in order to destroy the new population either a greater concentration of the same AMA, or another type of AMA must be used. In this way, together with gradual dilution and inactivation, commonly used solutions of sanitizers in which mops, lavatory brushes, cloths, etc. are kept, can become sources of organisms and be the means by which they are spread, thus completely defeating the object of sanitization.

### Growth inhibitory activity

The inhibitory action of any AMA against suitable target organisms is influenced by the pH, the presence of organic matter which through chemical reaction or absorption can reduce AMA action, the number of viable cells present, the stability of the agent under the conditions of use, the temperature, and the effect of dilution.

Growth inhibitory activity can be demonstrated in experiments which permit continuous contact between the test organisms and the agent. This can either be done by exposing the organisms to serial dilutions of the agent in liquid (see Figure 43) or solid media (see Figures 41 and 42), followed by incubation of the systems. Inhibition is shown by lack of visible growth in the liquid media, or by absence of growth in the near vicinity of the agents in solid media. Such results have to be obtained under carefully controlled conditions and not given too liberal an interpretation. A test as set up in Figure 43 shows that micro-organisms can grow in the presence of an AMA if the cell concentration is sufficiently high. The minimum inhibitory concentration (the MIC) is that concentration of AMA which prohibits detectable growth in the time period of the test. In practical terms the time period set will be appropriate to the conditions under which the AMA is used.

### Lethal activity

This is the ability of chemicals to kill micro-organisms in a short period of contact between the agent and the exposed cells. It, too, is

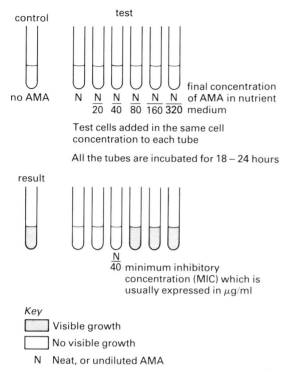

Figure 43  *Demonstrating the inhibitory activity of an anti-microbial agent*

affected by environmental and other factors – the period of contact, the concentration of the agent, the temperature, the number of cells present, the pH and the organic matter.

It is, therefore, assessed by varying the *time of contact*, and counting the number of surviving cells at various AMA concentrations. Within a limited period of time – up to 15 minutes – the longer the exposure time to a sanitizing or disinfecting agent, the greater its destructive effect. This can be demonstrated in the laboratory (see Figure 44). It is important to neutralize the AMA in the subsamples taken, otherwise it could continue to kill after the time period of contact. Beyond a time period of about 10–15 minutes, increases in exposure time are ineffective because those organisms which can resist the concentration of AMA may remain viable or may in fact continue to grow and reproduce. In sanitizing practice, for example, the time of

exposure may be only 10 minutes because operatives do not have unlimited time in which to carry out sanitizing tasks. So the concentration at which a sanitizing agent is used is in part determined by the time in which it is required to achieve a certain level of kill.

Use of the correct *concentration* for the elimination or control of target organisms is very important. As previously explained, use of a concentration which does not destroy (or inhibit) a high proportion of the target cells can result in the survival and multiplication of those which are more resistant. Equally, use of a concentration of agent which is greater than that required to kill the most resistant of the target organisms, will not achieve a more effective result – it is just wasteful. The experiment outlined in Figure 44 can be expanded by testing different concentrations in the same way. It would be found that as the concentration varied so did the time taken to achieve the same lethal effect on target organisms.

All AMA operate optimally at a particular *temperature*. Higher temperatures often help, but only if the AMA does not dissociate or become inactivated.

If a 99 per cent kill is achieved the effectiveness of that result is measured by the actual *number* of cells which the remaining 1 per cent represents. Thus, materials which before treatment have very high counts, are much more difficult to treat than those which have a low initial cell count.

The *pH* at which an agent is required to operate influences the result achieved (see Table 7).

The presence of *inactivating* materials such as proteins in blood, hardness in water, certain detergents, soaps and other substances can inactivate or reduce the effect of an AMA. A major problem in the food industry is that many sanitizers are liable to become inactivated by waste food and other organic matter. The greater the quantity of this *soil* the greater is the degree of inactivation of an applied sanitizer. In order to achieve adequate sanitization, organic soil must first be removed. This is very important when disinfectants are used, both in

*Part 1*

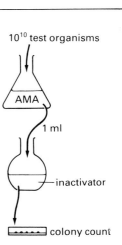

1   A standardized number (for example $10^{10}$) of test organisms in 1 ml is
    added to a flask containing 99 ml of test AMA at a certain concentration and
    temperature, giving $10^8$ organisms per ml.

2   At timed intervals – 30 seconds, 1, 2, 5 and 10 minutes – 1 ml samples are
    removed and added to another solution which neutralizes the action of
    the AMA.

3   Samples of the viable organisms are removed from this and the number of
    survivors are counted probably using a colony count technique.

4   *Results:* a time value will be found at which no viable organisms are
    recoverable.

*Part 2*

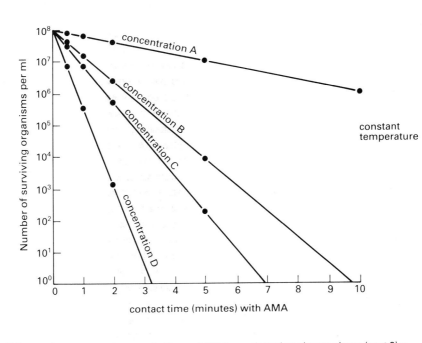

*Note:* If the results on several concentrations of AMA are plotted as shown above (part 2) a
suitable concentration of AMA can be found to destroy the population. Concentration A would not
be appropriate because in the maximum time available (10 minutes) an insufficient proportion of
cells have been destroyed. Concentrations B, C and D are satisfactory and choice would be made
on the practicalities of the time period of exposure. This type of test is known as a suspension test
and can be applied as a use-dilution test (see Chapter 10).

Figure 44   *Demonstrating the relationship between contact time and lethal effect of an AMA*

Table 7

| Agent | Optimum operating pH | Comments |
|-------|----------------------|----------|
| Hypochlorite | Near pH 7 | Corrosive at neutrality, and is therefore often used under alkaline conditions (e.g. at around pH 9.0) with less effect but also with less corrosiveness – important when metals are being treated |
| Iodophors | Acid conditions | They are more effective than chlorine below pH 5.0 |
| Quaternary ammonium compounds | Alkaline conditions | |

the cleaning of food handling equipment and in their use elsewhere. A test – the *capacity test* – has been devised to check by how much a sanitizer can be contaminated with organic soil before it loses its effectiveness (see page 151).

## Ionizing irradiation

The radiations which have an adverse effect on cells are those with wavelengths short enough to disrupt intra-cellular molecular structure, leading to chemical changes and alterations in cell metabolism. After slight exposure a damaged cell may be able to repair the damage and survive and possibly give rise to offspring of a new type known as *mutants*. Greater exposure results in more severe molecular damage which prevents reproduction and leads to cell death. These effects occur in all types of irradiated cell including those of animals and microbes.

*X-rays* (wavelengths 14 to 0·005 nm) and γ-rays (0·14 to 0·0005 nm) can penetrate deeply into solids and liquids, and split complex molecules into molecular parts known as 'ions' and for this reason are described as 'ionizing radiations'. The ions regroup into new molecules which may not function in the same way as those from which they originated. *Ultra-violet (UV) light rays*, although not as penetrating as X-rays and γ-rays, are strongly absorbed by exposed nucleic acids and protein molecules which again give rise to incapacitating chemical reorganizations (see page 128). *Visible light* is only weakly penetrating, yet can cause damage to molecules and have mild germicidal action, although its effects are minor in comparison with the other radiations mentioned (see Figure 45).

In any cell population the cells are able to resist exposure to irradiation to different degrees – some cells being more sensitive than others. The damaging effect which occurs depends on the dosage of irradiation absorbed – the greater the dosage the more damage occurs. One danger associated with the use of irradiation as a means of killing micro-organisms is the possibility that minimal dosages of irradiation which destroy only the more sensitive cells may lead to a build up of a radiation resistant population which requires greater exposure to eliminate. But if proper sanitizing procedures are also used, the build up of such populations should not occur.

Ionizing radiation dosages are measured in *rads* and *krads*. Experimentally the sensitivities of bacterial species have most frequently been measured in phosphate buffer. Such experiments have shown that wide variations in the dosages required to reduce each population to one-tenth of its original value occur between species. The dosage for the most sensitive bacteria in these experimental conditions is 2–3 krad, for the most resistant it is 110–120 krad. The exceptionally radio resistant species, *Micrococcus radiodurans* requires a much bigger dose – about 800 krad to decimate its population.

In sensitive species the death rate in the population is exponential. Thus, if the dosage to reduce the population to 10 per cent of the original is 21 krad then to reduce it to 1 per cent of the original will be 42 krad (2 × 21 krad) and to 0.1 per cent 63 krad (3 × 21 krad). However, in resistant species a considerably greater initial dose may be needed to decimate the population than is subsequently needed to further reduce the population from 10 per cent to 1 per cent of the original. Thus irradiation can have a selective effect on the micro-flora of the exposed food – selectively reducing the sensitive species to insignificant numbers and leaving the more resistant ones.

The most sensitive bacteria are gram negative rods, including *E. coli, Yersinia enterocolitica, Salmonellae* and *Shigellae* and members of the genera *Pseudomonas, Serratia* and *Proteus*. Raw meat and fish products tend to carry these types of organisms among their flora. Therefore when they are experimentally irradiated their subsequent spoilage would be due to the growth of the more resistant surviving species, which tend to be the psychotrophic *Acinetobacter-Moraxella* group of organisms.

Gram positive organisms show a range of sensitivities to ionizing radiations, some being sensitive and some being very resistant. It is really a matter of the combination of species present and environmental circumstances which determines the nature of the surviving population. Dry conditions tend to make all species more resistant than they would be under moist conditions. Dry bacterial spores are usually very radio resistant, and important in the food industry are the spores of *Clostridium botulinum*.

Other environmental factors influence susceptibility – the presence of oxygen, for example, enhances lethal effects, as do raised temperatures. Conversely, freezing protects cells. Yeasts respond to ionizing radiations in a similar manner to bacteria, but the kinetics of death depend on whether the cells are haploid (single chromosome) or diploid (paired chromosomes).

In the use of ionizing radiation as a preservative treatment for food, public health risks would be associated with those species which can survive low dose treatments, for example spores of *Clostridia* and *Bacillus*, provided they could grow subsequently. This risk could usually be overcome in practice by the combination of treatments of irradiation and heat, or irradiation and chemicals (such as sodium nitrate or sodium chloride), or by suitable post irradiation storage.

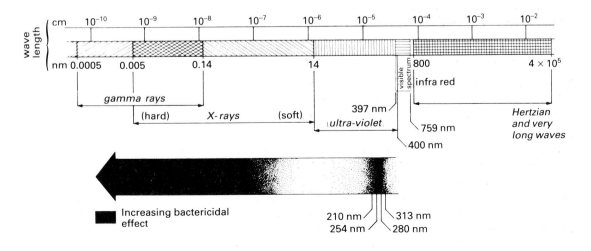

Figure 45 *Electromagnetic spectrum*

## Ultra-violet light

Ultra-violet (UV) wavelengths are longer than those of X-rays and have a range of 14–400 nm. UV rays of wavelength 300–400 nm have a mildly biocidal effect. These are the wavelengths occurring in sunlight which can penetrate the earth's atmosphere. All those of wavelength below about 290 nm are screened out and do not reach ground level. If they did, it would be with disastrous consequences to life.

UV lamps, designed for the purpose, can emit wavelengths in the most lethal range of around 260 nm and are used for killing micro-organisms in air or in liquids (such as water, or in heat sensitive pharmaceutical preparations). In experimental conditions, the characteristics of death are similar to those of ionizing radiations. The UV rays are absorbed by the intracellular RNA and DNA resulting in cell death; or, if the cells survive, resulting in an increased frequency of mutation. But because UV rays are not very penetrating, unlike ionizing radiations, they are most effectively used either for sterilization of surfaces or of *thin* liquid films.

The effectiveness of UV rays on a suspension of cells depends on the dosage received by those cells. The dosage is dependant on the intensity of the incident light and the duration of exposure. The intensity, which diminishes in relation to the distance from the lamp, is measured in $\mu W/cm^2$.

Different organisms vary in their susceptibility to UV light. For example, D values for *Pseudomonas aeruginosa* in saline solutions have been shown to be approximately one-sixtieth of those for those for *Micrococcus radiodurans* under the same conditions. Other examples are shown in Table 8.

Sometimes cells damaged but not killed by UV rays can later be reactivated by visible light – but research has shown that this has no practical significance in UV sterilization.

However, the destructive effect of UV light in sunlight on micro-organisms does have practical significance in the environment. The microbiological pollution of natural waters by coliforms and other pathogens in sewage disposed of into them, is reduced when the sunlight is bright. This effect is only significant in clear waters and not in the depths of the water or where it is opaque due to suspended matter.

Table 8

| | Organism | Time (seconds) to reduce the population to one-tenth of its original number* | |
|---|---|---|---|
| ATCC13939 | *Micrococcus radiodurans* | 1995 | |
| ATCC6538 | *Staphylococcus aureus* | 53.7 | |
| ATCC10541 | *Streptococcus faecalis* | 120.3 | (Intensity 100 |
| ATCC10231 | *Candida albicans* | 438.5 | $\mu W/cm^2$ – cells |
| ATCC8737 | *E. coli* | 81.6 | exposed in saline |
| ATCC19146 | *Pseudomonas diminuta* | 73.8 | solution) |
| ATCC8071 | *Pseudomonas putrifaciens* | 88.8 | |

*Adapted from R. L. Abshire, and H. Dunton, *Applied and Environmental Microbiology* no. 41, pp. 1419–23 (1981)

## Microwave energy

The use of microwave generating appliances to cook food is becoming more widespread both in industry and in the home. Microwaves are part of the electromagnetic spectrum – in the zone of radio frequencies. The frequencies of either $2450 \pm 50$ MHz or $915 \pm 25$ MHz have, by international agreement, been set aside for cooking and food processing. The microwaves can be generated in specially designed cookers from electrical energy.

The cookers for home use are very similar to conventional ovens and cook the food in batches; industrially they can be used for batch and continuous processing. Microwaves are absorbed by certain solids and liquids and not by others. Organic materials (such as foods) and liquids absorb them by a process from which heat is generated and it is by this heat that the food is cooked and the micro-organisms destroyed. Thus, the heat sensitivity of the contained organisms, the temperatures achieved, and the duration of exposure are what determines the percentages of organisms destroyed. Microwaves are reflected by metals (which are used to contain the waves inside the ovens) and pass through most glass, china and plastics. These materials are suitable for containers in which to microwave cook foods at home. However, before large scale food processing by microwave heating can occur, a suitable packaging material which allows the passage of microwaves and yet prevents absorption of water and oxygen while the product is stored, must be designed and to a large extent is still being sought.

Although microwaves cook by their heating effects their use in the cooking of foods and as a means of pasteurizing or sterilizing foods have to be treated differently from conventional applications of heat.

The time required to reach a particular temperature inside the food is influenced by several factors:

1  The initial temperature of the food – the lower it is the longer it takes to raise the food temperature to a predetermined level.

2  Food thickness – the greatest penetration of microwaves occurs in the outside 35–45 mm. Areas of food deeper than this will heat primarily by conduction of heat from the hotter outside layers. It has been found, for example that cooking food in thin portions in plastic pouches gives very rapid heat transfer and is the most satisfactory.

3  Distribution of moisture – this in 'solid' foods varies within the food and in practice leads to local 'hot spots' in the moist areas.

4  Food density – very dense foods require more energy input to heat them than less dense foods do. Foods of varying densities, for instance, composite foods such as cakes, stews, etc. therefore experience patchy heat distribution directly related to the distribution of the varied densities of the food components.

Good cooking practice overcomes the latter problem by advising turning or stirring the food or rotating a joint, as appropriate. Improvements in the design of ovens will gradually help to overcome these problems. Further, the practice of allowing food cooked by a microwave oven to stand for a while after exposure, in order for heat to pass through by conduction, should, *if* the temperatures are high enough, have advantageous effects both in terms of eating qualities and microbiological safety.

Industrial applications of microwaves are being developed today to determine the feasibility of their use. For example, trials on the *sterilization* of cream and orange juice have been reported; experiments to pasteurize cured hams, milk, soya bean curd and oysters have also been undertaken – all with the object of limiting microbiological deterioration (as with conventional heating methods) and improving nutritional and/or textural properties.

*Dehydration* by microwave is now possible, for example pasta, onions, tomato paste, fungal protein, bacon. Other applications in the use of microwaves include freeze drying and vacuum drying, where in some processes, improvements over conventional methods have been reported. Finally, microwaves are of great assistance in the rapid partial thawing of deep frozen meat (−18°C) prior to butchering (−3 to −2°C).

Microbiological problems may arise if, through uneven heating, patches of food are underexposed to lethal heat. In these 'pockets' bacteria may sometimes survive. Failure to observe good hygienic practices may permit these survivors to either grow or survive in the food after cooking. Experiments have shown that this can happen.

# Chapter 4

# Food poisoning

Food poisoning and food infections are fairly common occurrences that most of us will experience at some time. Although food poisoning may be caused in a variety of ways, it is generally an acute condition. The symptoms of illness arise within hours, or at the longest, days after the consumption of the affected food. It usually manifests itself as an attack of gastroenteritis the symptoms of which are nausea, vomiting, pyrexia, diarrhoea and abdominal pain. In some types of food poisoning there may also be symptoms indicating that the nervous system is affected.

Foods cause food poisoning when:

They are contaminated with poisonous chemicals
They are poisonous themselves
They contain food poisoning micro-organisms and/or their toxins

## Food contamination by poisonous chemicals

Acute poisoning results from consuming food excessively contaminated with heavy metals, examples being antimony, copper, cadmium, lead or zinc.

The maximum levels of heavy metals permitted in food, are controlled by law. Even so, accidents and unforeseen circumstances arise. For example, the use of lead caps to crown wine bottles carries the risk of more than the permitted level of lead in the wine. A case was reported in Great Britain in 1981 involving canned grapefruit which proved to have danger-ously high levels of tin and lead causing metal poisoning in a woman consumer.

In addition to acute poisoning, chronic poisoning may be caused through the accumulation in the body of metals occurring in food. The importance of monitoring the levels of heavy metals permitted in food is reflected by the continuous surveillance undertaken by government agencies who report regularly and warn of dangers when necessary.*

It is a well known fact that direct consumption of pesticides and herbicides is extremely dangerous and can, in some cases, cause death. Considerable concern is being expressed at the unknown long-term effects of consuming foods contaminated with pesticide residues. Legal control over their levels in foods is exercised both through EEC and UK legislation.

Other organic materials can be extremely toxic as was illustrated by the tragic case in Spain in 1980 where many hundreds of people were acutely and chronically ill through eating chemically contaminated olive oil. Establishing the cause has proved difficult but there is evidence that the food oil was contaminated with industrial oils which had been marked with aniline dyes to distinguish them from edible oils.

Materials which may come into contact with food are also controlled by law (The Materials

---

* *Evaluation of certain food additives and the contaminants, mercury lead and cadmium*, 'WHO Technical Report Series', no. 505 (WHO 1980); and *Survey of copper and zinc in food* (HMSO 1980).

and Articles in Contact with Food Regulations, 1979), so the use of potentially or actually dangerous substances such as asbestos, is very strictly controlled.

## Naturally poisonous plants and animals

There are a number of wild and garden plants which, if eaten in sufficient quantities, cause illness. Examples include: henbane, deadly nightshade, laurel leaves, hemlock and laburnum seeds. Fungi are often, with justification, regarded with some suspicion. In Britain, of the many types of poisonous 'toadstool', the *fly agaric* is perhaps the best known. The condition *ergotism* is a poisoning due to eating cereals (usually as flour in bread) infected with the fungus *Claviceps purpurea*. Rye is the principal cereal affected, but because rye is not eaten very much today the incidence of ergotism is very low. Herbs, both wild and cultivated, are used as food flavourings, and some, in addition, have medicinal uses. The chemicals which are of value medicinally can be poisonous if taken in excessive doses. *Digitalis* which is extracted from the foxglove has medical uses yet it is also toxic.

Some plants cultivated for food can also be poisonous if not prepared properly. The leaves of rhubarb contain oxalic acid. Hydrogen cyanide can be released from the kernels of most stone fruits, for example apricots and almonds. A case was reported in Israel in 1981 where twenty-four children were poisoned and four died as a result of eating apricot kernels. Prolonged boiling makes the fruit safe to eat. Beans of many varieties, for example Lima beans and red kidney beans, contain toxic substances which can be effectively removed by prolonging soaking, followed by boiling for at least ten minutes prior to use.

Some animal tissues are naturally toxic and will cause illness if consumed. Some animals store vitamin A in their livers in quantities not tolerated by man, for example the polar bear. Fortunately, for most of us, the likelihood of eating liver from this source is not very great.

Sea foods of various types have proved to be toxic because they have retained the toxin produced by certain protozoan dinoflagellates occurring in plankton (for example, the dinoflagellate *Gonyaulax tamarensis*). Mussels have occasionally caused poisoning this way. Fish from tropical and subtropical waters can cause *ciguatera* poisoning in man. This appears to be due to a poison originating in a dinoflagellate which is transmitted to man via the food chain. The toxin is tasteless, odourless, and is not destroyed by cooking or gastric juices.

*Scombroid* poisoning is caused by eating the slightly spoiled flesh of certain dark-fleshed fish such as mackerel and tuna, and also, on occasions, canned sardines and pilchards. The flesh of the fresh fish does not cause poisoning as some degree of spoilage is necessary. Contaminating bacteria grow on the dark flesh of scombroid fish which is rich in the amino acid histidine. Samples of fish suspected of having caused poisoning have been shown to contain high levels of histamine, and the symptoms of poisoning include those of a histamine reaction, that is, nausea, vomiting, diarrhoea, headache, rashes, dizziness and respiratory distress. Cases of poisoning from this source generally arise from eating smoked fish, where the peppery flavour change associated with the chemical changes is perhaps masked by the smoked flavour. Scombroid poisoning is generally rare in Britain but four cases were reported in 1978 and 196 in 1979.

## Microbial food poisoning

Microbial food poisoning is caused either by the ingestion of food contaminated with large numbers of pathogenic organisms or by the ingestion of toxins produced by micro-organisms within the food before its consumption.

The main organisms known to be associated with food poisoning are:

1   Organisms of the genus Salmonella
2   *Clostridium perfringens*
3   *Staphylococcus aureus*
4   *Bacillus cereus*
5   *Clostridium botulinum*

6 Species of the genus Campylobacter
7 Enteropathogenic *E. coli*
8 *Streptococci*
9 *Vibrio parahaemolyticus*
10 *Yersinia enterocolitica*
11 Mycotoxic fungi

The order in which these are listed is not significant, except that salmonellae, *Clostridium perfringens* and *Staphylococcus aureus* have for many years been regarded in Britain as the organisms which together cause the most outbreaks of food poisoning – salmonellae causing the greatest number. The situation is gradually changing in the sense that although salmonellae still cause the greatest number of incidents, other organisms are now identified as also causing food poisoning, and are isolated from samples of suspected foods with regularity. For example, the first reported case of food poisoning identified as due to *Bacillus cereus* was in 1971, although outbreaks had been reported in Europe before that time. Since 1971 a small but significant number of cases have been reported every year. In fact bacterial food poisoning has been a notifiable disease in England and Wales since 1938. A large number of people suffer every year from gastrointestinal upsets as a result of eating contaminated food. Different-sized groups of people may be affected. The terms used to describe these are shown in Table 9. The illness is very unpleasant and could largely be avoided if there was more understanding of how food poisoning organisms spread. Food poisoning hits the already debilitated – the old and the very young – hardest, and every year, sadly, some people die of the illness. The number of incidents of food poisoning occurring annually over a ten year period is shown in Table 10.

One category of incident – the general outbreak – is more closely analysed in Table 11, which shows the relative responsibilities of the different bacteria.

Table 9 *Definition of terms associated with food poisoning*

| Term | Definition |
| --- | --- |
| A general outbreak | Two or more related cases or excretors in persons of different families |
| A family outbreak | Two or more related cases or excretors in persons in the same family |
| A sporadic case | One case which, as far as was ascertained, was unrelated to other cases |
| An incident | Any one of the three mentioned above |

*Source*: The Ministry of Health and Public Health Laboratory Service, *Monthly Bulletin*, HMSO 1980.

## Salmonella species

### Morphology

Organisms of the *salmonella* group are gram negative facultative bacilli which are, with some exceptions, motile. They do not form spores but

Table 10 *Incidents of food poisoning and salmonellosis in England and Wales 1970–9*

| | 1970 | 1971 | 1972 | 1973 | 1974 | 1975 | 1976 | 1977 | 1978 | 1979 |
| --- | --- | --- | --- | --- | --- | --- | --- | --- | --- | --- |
| General outbreaks | 194 | 207 | 153 | 186 | 208 | 247 | 245 | 223 | 214 | 217 |
| Family outbreaks | 641 | 545 | 473 | 603 | 383 | 328 | 372 | 328 | 365 | 325 |
| Sporadic cases | 4455 | 4941 | 3469 | 4729 | 3883 | 6501 | 5092 | 5273 | 6835 | 7413 |
| All incidents | 5290 | 5693 | 4095 | 5518 | 4474 | 7076 | 5709 | 5824 | 7414 | 7955 |

*Source*: *British Medical Journal*, no. 281, 817 (1980).

can withstand conditions such as drying and freezing for considerable periods.

As can be seen from Tables 11 and 12 salmonellae are responsible for the majority of food poisoning outbreaks and cases in England and Wales. There are over 1600 strains (or 'serotypes') of salmonellae known, but the ones isolated from humans after infection are limited to a small number of types.

### Serotypes

Serotyping is a method of identifying an organism through the chemical patterns on its surface. It has been found that closely related strains of organisms differ, sometimes in minute ways, in their surface structures.

When the bacteria invade or are introduced into a suitable warm-blooded animal, the animal's defence system responds to fight the infection by producing *antibodies* (AB) specific to the invading bacteria. These antibodies circulate in the blood and can be separated in the *serum* (hence *serology, serotypes*). It is the surface structures of the micro-organisms – the *antigens* (AG) – to which the antibodies are specific.

Laboratory techniques have been developed using specially prepared *antisera* (serum containing antibodies) by which unidentified bacteria (for example, salmonella, isolated from a patient with food poisoning) can be specifically identified by testing them against a range of different but known antisera. Compatibility can be observed by a visible reaction between AG and AB. Using a logical sequence of tests, the organism can be precisely identified which is important in tracing the epidemiology of salmonella infections.

The commonest serotype associated with human infection isolated in this country is *Salmonella typhimurium*, although there has been a reduction in the percentage of incidents caused by this organism in recent years (see Figure 46). Figures show that *Salmonella typhimurium* accounted for 80 per cent of the serotypes isolated from incidents in 1955 but by 1969 this figure had fallen to 36 per cent. The order of predominance of serotypes changes from year to year. Between 1975 and 1979 the number of general outbreaks of salmonella food poisoning caused by *Salmonella typhimurium* fell to 28 per cent, but those caused by other serotypes increased, for example *Salmonella agona, Salmonella enteritidis, Salmonella heidelberg, Salmonella virchow* and *Salmonella hadar*. *Salmonella typhimurium* is the salmonella serotype largely responsible for infections in cattle whereas the other serotypes mentioned are derived mainly from chickens and turkeys.

The number of general outbreaks of food

Table 11   *General outbreaks of food poisoning and salmonellosis in England and Wales 1970–9*

| Organism | 1970 | 1971 | 1972 | 1973 | 1974 | 1975 | 1976 | 1977 | 1978 | 1979 |
|---|---|---|---|---|---|---|---|---|---|---|
| Salmonellae | 138 | 163 | 103 | 128 | 145 | 134 | 124 | 126 | 143 | 138 |
| *Clostridium perfringens* | 31 | 29 | 29 | 30 | 42 | 69 | 84 | 78 | 38 | 53 |
| *Staphylococcus aureus* | 18 | 8 | 8 | 6 | 7 | 15 | 13 | 6 | 11 | 14 |
| *Bacillus cereus* | 0 | 2 | 4 | 16 | 10 | 20 | 13 | 8 | 10 | 4 |
| Other/unknown | 7 | 5 | 9 | 6 | 4 | 9 | 11 | 5 | 12 | 8 |
| Total | 194 | 207 | 153 | 186 | 208 | 247 | 245 | 223 | 214 | 217 |

*Source: British Medical Journal*, no. 281, 817 (1980).

poisoning caused by salmonellae has been rising over recent years. *Salmonella typhimurium* is still the commonest serotype isolated, but *Salmonella hadar* has gradually come into prominence, increasing from 31 isolations in 1971, to 2480 in 1979. The probable reason for the change in prominence is the alteration in national eating habits brought about by the rise in the price of beef which has led to a decrease in the amount consumed. Chicken and turkey on the other hand are the cheapest and most popular forms of animal protein in the UK.

### Mode of infection and growth characteristics

The organisms multiply in food under suitable environmental conditions of temperature (37°C optimum) and moisture. In order to cause infection varying numbers must be ingested – as many as $10^6$ for some species of Salmonella, as few as several hundred cells for others. Pas-

Table 12   *Cases of bacterial food poisoning and salmonella infection\* according to year and agent†*

| Organisms | Year | | | | | | | | | |
|---|---|---|---|---|---|---|---|---|---|---|
| | 1970 | 1971 | 1972 | 1973 | 1974 | 1975 | 1976 | 1977 | 1978 | 1979 |
| *Salmonella* | 6848 | 6784 | 4849 | 6871 | 5663 | 8894 | 7465 | 6501 | 9086 | 9912 |
| (per cent) | (79) | (84) | (81) | (80) | (66) | (74) | (68) | (71) | (86) | (83) |
| *Clostridium perfringens* | 1263 | 978 | 1026 | 1311 | 2767 | 2418 | 2924 | 2576 | 1042 | 1607 |
| (per cent) | (15) | (12) | (17) | (15) | (32) | (20) | (27) | (28) | (10) | (10) |
| *Staphylococcus aureus* | 523 | 302 | 116 | 168 | 103 | 443 | 510 | 81 | 301 | 328 |
| (per cent) | (6) | (4) | (2) | (2) | (1) | (4) | (5) | (1) | (3) | (3) |
| *Bacillus cereus* | – | 15 | 16 | 66 | 51 | 82 | 54 | 37 | 143 | 22 |
| (per cent) | (–) | (–) | (–) | (1) | (1) | (1) | (–) | (–) | (1) | (–) |
| *Vibrio parahaemolyticus* | – | – | 13 | 17 | 5 | 96 | 44 | 9 | 9 | 6 |
| (per cent) | (–) | (–) | (–) | (–) | (–) | (1) | (–) | (–) | (–) | (–) |
| Other organisms | – | – | – | 141‡§ | – | 10‡§ | – | – | 9§‖ | 6 |
| (per cent) | (–) | (–) | (–) | (2) | (–) | (–) | (–) | (–) | (–) | (–) |
| Total bacterial causes | 8634 | 8079 | 6020 | 8574 | 8591 | 11 943 | 10 997 | 9204 | 10 590 | 11 881 |
| (per cent) | (100) | (100) | (100) | (100) | (100) | (100) | (100) | (100) | (100) | (100) |

\*Excluding symptomless excretors of salmonellae
†In addition, several outbreaks of other or unknown aetiology were reported each year
‡*E. coli*
§*Bacillus* spp.
‖*Clostridium botulinum*

*Source*: E. Hepner, *Public Health*, no. 94, pp. 337–49 (1980).

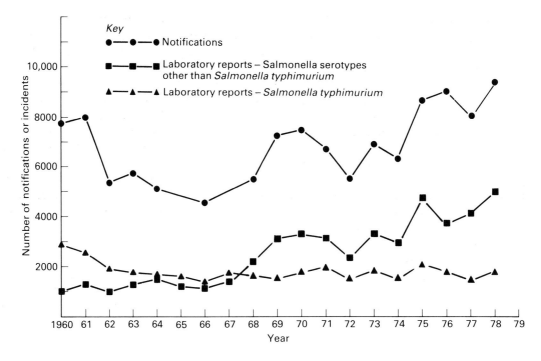

Figure 46   *Incidents of food poisoning and salmonellosis between 1960 and 1979 in England and Wales*
*Source*: P. Forbes, N. S. Galbraith and R. T. Mayon-White, *British Medical Journal*, No. 281, pp. 546–9 (1980)

teurization temperatures such as those used for the treatment of milk are sufficient to destroy the vegetative cells.

### Effect on consumer

Salmonella food poisoning has an incubation period of between 6 and 48 hours, but it is usually apparent within 12 to 24 hours. The main symptoms are abdominal pain, diarrhoea, prostration and frequent vomiting. Fever is nearly always present and the illness lasts from 1 to 8 days, sometimes longer. The majority of incidents are not fatal but occasionally death from salmonellosis occurs, usually in the elderly or in people debilitated by another condition.

### Habitat

Salmonella organisms are widely distributed in nature, their common habitat being the intestinal tract of mammals, birds and reptiles. They are excreted in faeces and in this way may contaminate any objects with which the faeces

come into contact. The sources of the organisms are considered below.

### Man as a source

Man excretes salmonellae in his faeces after infection with the organisms but may or may not show symptoms of food poisoning. Those who suffer the infection excrete the organisms in large numbers during illness, and in decreasing numbers during convalescence, and are known as *acute* and *convalescent* cases respectively. The period of infectiveness varies from a few days to weeks or even longer, (about 5 per cent of cases). The convalescent carrier state has, over the years, gradually extended in length with the rise of the newer serotypes. For example, in the case of *Salmonella agona*, the convalescent carrier state lasts, on average, 18 weeks. Therapeutic antibiotic treatment for salmonella food poisoning has the effect of extending the time the patient excretes the organisms in his faeces, although special treatment with antibiotics to eliminate the organism and obviate the semi-permanent carrier state is normally effective.

*Ambulant* cases suffer mild symptoms which they may not associate with food poisoning, but nevertheless excrete the organisms. Individuals who pass the organisms, probably in small numbers without showing symptoms of infection, are called *healthy* carriers or symptomless excretors. A *chronic* carrier shows persistent discharge of the organisms over a long time in transitory bursts. In the developed areas of the world, where good sanitation and clean water supplies exist, the salmonella carrier rate is rarely higher than 0.3 per cent.

The organisms are excreted in the faeces and may in this way contaminate the hands. To encourage cleanliness and to discourage transfer of micro-organisms, notices stating 'Now wash your hands' are put in the toilets of establishments concerned with the preparation and sale of food to the public. (See Figure 47.)

### Animals as a source

Salmonellae are widely distributed among different species of animals including domesticated farm animals and household pets. They are also present in wild animals, rodents, insects and birds. There is always a proportion of these which act as carriers excreting salmonellae, so that anything which comes into contact with their faeces is liable to become infected. (see Figure 48.)

### Meat

Animals which are reared for meat are often excretors of salmonellae and are the major source of human salmonella infection in this

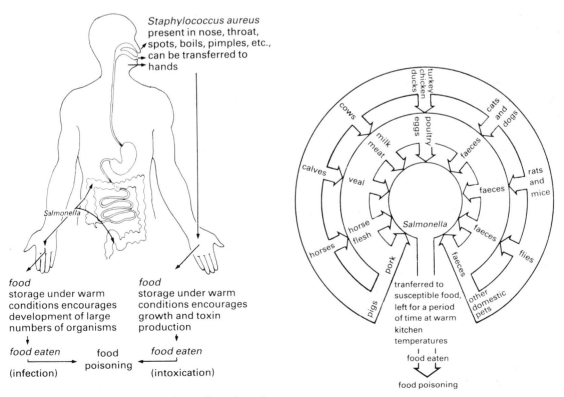

Figure 47   *Human sources of salmonella and staphylococcus*

Figure 48   *Animal sources of salmonella*

country. Animals, in the same way as humans may actually suffer the infection or become transient symptomless excretors. Certain species seem to carry the organisms to a much greater extent than others although the incidence also depends on the methods of rearing and feeding.

There is evidence that animals reared intensively have higher carrier rates than those reared non-intensively. In the intensive rearing of cattle each unit consists of a gathering of cattle from a large number of farms. If any of the animals present are infected, this may result in the infection of the whole unit.

Cattle (particularly calves), pigs and poultry are the major sources of salmonellae in meat in this country. Horses are also an important source in countries where horsemeat is eaten both by humans and pets. Sheep carry the organisms at a low rate in the UK, and consequently mutton and lamb are less frequently infected. The salmonellae spread from the adult animals to the young.

Surveys also show that although the rate of faecal infection in individual farms may be low, transportation of animals close to one another on long journeys, and stress in travel, market and lairage, increase the rate of excretion. By the time the animals have been marketed and mixed at the slaughterhouses, the infection rate may have risen to about 20 per cent before slaughter.

Calves which have been infected at the rearing farm and are then sold to other farms, where they will be fattened before slaughter, serve as a further means of spreading salmonellae.

The length of time in the lairage prior to slaughtering is now limited by legislation to a maximum of 72 hours so as to prevent cross-infection between animals. The amount of cross-infection in the first 12 hours is reported to be slight. This is illustrated in a survey carried out on pigs in a slaughterhouse which showed that infection of the mesenteric lymph nodes was 5–11 per cent in pigs kept for 1 day, but in those kept for longer periods, the figure rose to between 32 and 96 per cent. In another survey carried out on calves, isolation of *Salmonella typhimurium* was 0.6 per cent after 1 day in the slaughterhouse lairage and 36.6 per cent after 2.5 days.

Care must be taken in the slaughtering of animals as cross-contamination may occur in the slaughterhouse when the intestinal contents of infected animals may be passed on to healthy carcases and in this way get into the butcher's shop and finally into the kitchen. Salmonellae may also be derived from the animals' coats and hooves. Operatives, knives and any equipment used may serve as a means of cross-contamination.

*Animal feedstuffs and fertilizers*

Contaminated feedstuffs are known to be a means whereby animals such as cows, pigs and poultry either become transient symptomless excretors, or suffer actual illness. Feedstuffs are often imported from countries where sanitary conditions are inadequate and methods of production poor. Feeds are produced from animal, marine and vegetable sources and serve as supplies of proteins, vitamins and minerals. Investigations carried out in this country on feedstuffs of both plant and animal origin have shown the presence of a large number of different salmonella types. Feeds are produced in the form of meals and mashes, and are either used in that form, or are pelleted before distribution. Some kinds of meals are regarded as low grade and are used by farmers as soil fertilizers. Meals and mashes undergo a heat treatment during their production which is usually regarded as sufficient to kill salmonellae, but often the living organisms can still be isolated from the finished product. Therefore, either the heat treatment is inadequate, or there is post-treatment contamination. When meals and mashes are made into pellets they undergo additional heat treatment which helps to further reduce the number of organisms.

In one investigation in which salmonella-free pigs were fed with contaminated feed, the commonest serotype isolated from the mesenteric glands was *Salmonella typhimurium*, although this organism was present in the feed in lower numbers than other serotypes isolated, indicating its greater invasive power. The same investigation also showed that when given contaminated feed, animals will usually develop a latent infection but not a clinical infection. Several research workers in this field have suggested that if the raw materials were sterilized by heat, or another means, such as irradiation, this source of infection could be eliminated. New processing plants are being designed to provide improved methods of heat treatment.

*Poultry and eggs*
Poultry are likely excretors of salmonellae. The organisms are carried in the intestine, although in ducks the ovary may also be infected. In the mass production of poultry, especially in broiler units, large numbers of prepared carcasses may be infected from a low rate of excretion in the living birds, as a result of cross-contamination in processing (see page 172). The same equipment is used for dealing with many birds, so if one carcass is infected, the organisms may be passed to a large number of others. The use of chlorine in the washing water reduces the spoilage flora but does not eliminate salmonellae. A much higher percentage of carcasses leaving a large processing unit may therefore be contaminated than the percentage of infected birds entering the unit. Legislation now controls the processing of poultry – the Poultry (Hygiene) Regulations, 1976. The object of such legislation is to reduce the spread of salmonellae as well as to improve general hygiene standards.

The poultry industry in this country is arranged in such a way that the eggs and the chicks are produced in a limited number of centres whose responsibility it is to maintain the genetic characteristics of the stock. Infection of the master stock may result in the distribution of salmonellae to rearing farms and producers throughout the country.

Whole eggs may be contaminated on their outer surface with faecal material which may contain salmonellae. These organisms penetrate the shell or enter through cracks in it. Such eggs are of particular concern in the preparation of liquid bulk egg since one egg is capable of contaminating the mass. The routine pasteuriza-

tion of liquid egg is required by legislation brought into effect on 1 January 1963 (see Appendix) and has served to eliminate this risk.

Ducks are more susceptible to salmonella infection of the ovary, and their eggs may therefore be laid already containing the organisms. Even if the ovary is not infected, duck eggs may become infected because faecal material can penetrate the shell in the same way as for hen eggs. The tendency of ducks to lay their eggs in damp places increases this possibility, and these eggs should therefore only be used in food which is cooked thoroughly or if eaten as such, should be boiled for ten minutes.

*Pet food*
A large number of strains of salmonellae, many of which are associated with human infections, have been isolated from both imported and home-produced pet meat. In one outbreak, reported in 1967, two families who did not know each other but who lived in the same town both suffered an attack of food poisoning. Salmonellae of the same type were isolated from both families and the only common factor was the pet shop from which they obtained their pet meat and, in fact, salmonellae of the same serotype were also isolated, from the pet shop.

In an attempt to eliminate pet food as a source of infection the Meat (Sterilization) Regulations, 1969 were introduced. They require 'all knacker meat and meat (other than the meat of a rabbit or hare) which is imported otherwise than for human consumption, as well as all butchers' meat or imported meat which in either case is unfit for human consumption, to be sterilized before entering the chain of distribution'. Under these regulations, horse meat which has been passed as fit for human consumption does not have to be sterilized when used as pet food. It was hoped that these regulations would eliminate these sources of infection and the degree of success is currently being assessed through discussions at the Ministry of Agriculture, Fisheries and Food.

*Milk and dairy products*
Raw milk and cream may contain salmonellae

derived from faecal material, hides and hooves of the producing animals. In one incident involving 700 people who had consumed raw milk, *Salmonella dublin* was isolated from 190 of them. In an investigation in the USA in 1979 raw milk contaminated with *Salmonella dublin* had caused salmonella infections in people who had consumed the milk. In this particular case it was shown that 31 cows in the herd of 3100 were excretors.

Any food, dish or commodity made from raw milk products, as well as the raw products themselves, could therefore be contaminated with salmonellae; pasteurization of milk and cream obviates this risk.

### Fish and sea foods

Fish may be regarded as more or less free from salmonellae as these organisms do not form part of the normal gut flora of marine fish, or of fish caught in unpolluted waters. However, sewage pollution could contaminate the fish.

### Seagulls

It has been reported that 13 per cent of seagulls carry salmonellae, picked up, it is believed, from the places where they feed – refuse tips, sewage works, etc. They are thought to be an important source for the spread of salmonellosis by contaminating fields and pastures with their droppings and so effecting the transmission of the disease to farm livestock.

### Wild birds

In a manner similar to seagulls, other wild birds such as sparrows and starlings probably carry salmonellae. Two outbreaks of salmonellosis in the same psychiatric hospital in 1975 and 1979 indicated a possible association between the infection in some of the patients and wild birds in the kitchen. The kitchen was a large barn-like place with sparrows and other birds living and roosting on rafters, beams and windows. The probability was that contaminated droppings from the birds fell on to uncovered food. The organisms subsequently multiplied in the uncovered food and caused illness in the patients.

### Vegetables and fruits

Raw fruits and vegetables are potentially dangerous if they are sprinkled or washed with infected water or become in any way faecally contaminated.

### Salmonellae from other sources

Cats and dogs are known to excrete salmonellae in their faeces, and as they are pets they may be responsible for infection in man. This is also true of reptiles such tortoises and terrapins which are reputed to have been responsible for some incidents in children.

Rats and mice are also excretors, and measures to prevent their entry into premises should be taken. Rats and mice are victims of the environment – rats, especially, live in filthy surroundings such as sewers, and from these places they invade food premises. Rodents on

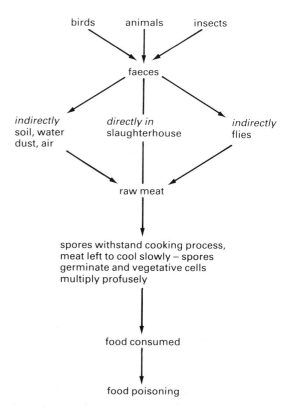

Figure 49    *Animal sources of* Clostridium perfringens

farms may eat contaminated animal feeds and may become excretors of salmonellae. So, besides the physical damage they are capable of doing, they are also able to spread food poisoning organisms.

Flies tend to lay their eggs in dung and faeces, or in refuse, and may fly from such places to open food. Improved environmental hygiene serves to reduce the number of breeding places for insect infestation. Insects may carry salmonellae in their intestine and contaminate food by their faeces or by their regurgitated food.

## Clostridium perfringens

### Morphology
The organism is a gram positive spore-bearing, square-ended bacillus.

### Habitat
*Clostridium perfringens* is carried in the bowels of animals, birds, insects and man. In the spore form it can survive for long periods in the air, dust, soil and water (see Figure 49).

If animals are subject to stress prior to slaughter, *Clostridium perfringens* can become systemic – reaching organs such as the liver, kidneys and muscle which are not normally affected.

There is a risk during the slaughter and carcass dressing that the contents of the viscera may contaminate the carcass. Spores and vegetative cells of *Clostridium perfringens* have been detected on the surface of commercial carcasses of meat – lamb, pork, beef and offal – and a high percentage of strains shown, by serotyping, to be food poisoning strains. Vegetables contaminated with soil may also bring *Clostridium perfringens* into the kitchen and into contact with food.

### Mode of infection and growth characteristics
When the vegetative cell is transmitted to a suitable food it multiplies rapidly (every 10 to 12 minutes) if the conditions are optimal. It thrives under anaerobic conditions and this gives some indication of the kind of food that may give rise to food poisoning – cooked meats such as stews,

pies and large joints. The organism multiplies in the food and, where large numbers are eaten, food poisoning can result. It is thought to be due to a toxin released in the intestines which is not present in the food before consumption. The optimum temperature for growth lies between 43 and 47°C, some growth occurs at 50°C but none at 55°C.

If the food is cooked at a high enough temperature, all the vegetative cells will be destroyed, but the spores can survive. If this food is not eaten immediately, but is allowed to cool slowly in a warm place, it will reach a temperature which is suitable for spore germination (around 50°C). Spores in fact require heat to initiate germination and this is provided by cooking (see Figure 50). Cooking also drives off oxygen so that anaerobic conditions are created which favour rapid growth.

The vegetative cells are readily destroyed by temperatures below 100°C. The spores of some strains of *Clostridium perfringens* are heat resistant and can withstand boiling for 1 to 4 hours. These spores will survive all cooking procedures except pressure cooking and industrial retorting. Outbreaks of food poisoning have been recorded which were due to heat sensitive strains of *Clostridium perfringens*. Such outbreaks are usually associated with circumstances similar to the following: when rolled joints of meat are used; when large joints are cooked from a frozen state, or when large joints are cooked and the time and temperature of cooking are inadequate to ensure that the temperature reached in the centre is sufficient to destroy either the vegetative cells or the spores. If cooking is then followed by a long, slow cooling period, the surviving spores will germinate and the vegetative cells which emerge may then multiply until they are present in sufficient numbers to cause food poisoning.

### Effect on consumer
The main symptoms of food poisoning due to *Clostridium perfringens* are abdominal pain, diarrhoea and prostration. There is a delay in the onset of symptoms of between 8 and 22 hours, but they usually appear within around 10

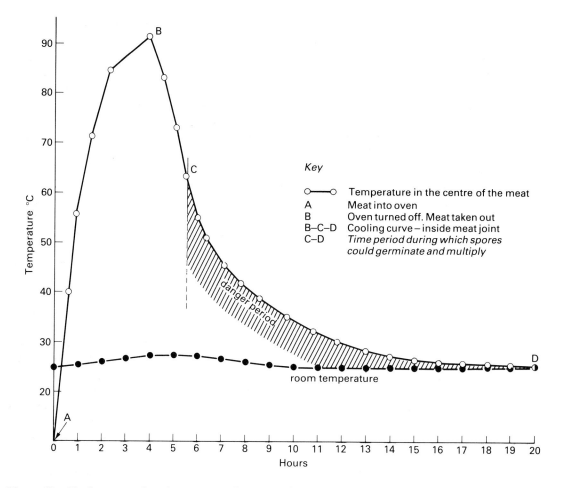

Figure 50   *The heating and cooling curve in the centre of a joint of meat which has been roasted and then left to cool at room temperature*

to 12 hours after consumption of the infected food and last for 12 to 24 hours.

## Staphylococcus aureus

The genus *Staphylococcus* is subdivided into two groups of strains – *Staphylococcus aureus* and *Staphylococcus epidermidis*. The primary habitat of the group *Staphylococcus aureus* is the skin and the naso-pharynx of man and warm-blooded animals. Potentially pathogenic, they can cause a wide range of infections and intoxications; some strains are responsible for causing food poisoning.

### Morphology
*Staphylococcus aureus* is a non-motile gram positive coccus. The organisms are arranged in clusters which are frequently compared with bunches of grapes.

### Habitat
Species of *Staphylococcus aureus* are carried in the noses of 30 to 50 per cent of the human population, and on the skin and hands of 14 to 44 per cent. It has been further estimated that 15 to 20 per cent of the human population carry enterotoxigenic staphylococci, that is, those species liable to cause food poisoning. *Staphylo-*

*coccus aureus* may be found in locations such as the nose, throat, groin, armpits and umbilicus without causing any discomfort, or it may be present in large numbers in inflamed lesions and skin infections such as whitlows, boils, pimples, acne, carbuncles, styes and barber's rash. Animals may also carry staphylococci. For example, cows suffer from mastitis due to staphylococcal infection of the udder and, as a result, the organisms may be present in raw milk and any of its by-products. In addition, surveys of food products show that *Staphylococcus aureus* can be widely isolated from fresh meats (chicken, pork, liver, spiced ham and others), fish, frozen meat by-products and a wide range of cooked and prepared foods. Thus food products may be primary sources of enterotoxigenic staphylococci.

### Mode of infection of food
The organism is nearly always passed to food from a human source (particularly by the hands) or by cross-contamination from another food, utensils, surfaces or equipment previously contaminated by humans, though it can also be contained in a raw food of animal origin (see Figure 47).

### Growth and environmental resistance
*Staphylococcus aureus* is a nutritionally demanding organism, but its growth requirements are supplied by most foods and it will grow provided the temperature, available water and pH permit. The growth range of *Staphylococcus aureus* is from 6.7 to 46°C, multiplication occurring slowly at low temperatures and the optimum temperature for growth being between 30 and 40°C. It is a facultative anaerobe which grows best in the presence of oxygen. It can grow within the pH range 4.5–9.3, with optimum growth between 7.0 and 7.5.

Staphylococci are generally considered to be among the more resistant of the non-sporing bacteria to environmental stresses. They possess the ability to grow in solutions of low water activity; they tolerate high concentrations of sodium chloride (up to 15 to 20 per cent – strain variable), and of bile (up to 40 per cent); and

they can be successfully freeze-dried. They tend to be resistant to drying, for viable cells can be isolated from dried products after variable periods of storage, yet their sensitivity to heat and radiations is not outstanding and is comparable to that of other vegetative bacteria. Pasteurization times and temperatures, similar to those applied for milk or egg, are sufficient to kill the vegetative cells.

### Enterotoxin production and destruction
Food poisoning is caused by the ingestion of the staphylococcal enterotoxins contained in food in which enterotoxigenic staphylococci have grown. In the UK staphylococcal food poisoning is responsible for only about 2 per cent of all reported cases, yet in other countries, for example, the USA and Hungary, it is the highest reported cause. Its incidence appears to be related to national dietary habits and it is possible that in the USA it correlates with the very high number of commercially prepared foods and communally eaten meals.

Strains of *Staphylococcus aureus* produce at least six enterotoxins – A, B, C, D, E and F. A survey has shown that strains producing enterotoxin A predominate among staphylococci isolated from foods.

Enterotoxins are produced during growth – with most produced in the logarithmic phase. Growth at 37°C allows a greater accumulation of toxin than growth at a low temperature. Enterotoxin A is produced in the log. phase and enterotoxin B in the stationary phase. The toxin accumulates in the food and cell levels of at least $10^6$/gm must be reached for the toxin concentration to be high enough to cause food poisoning (after eating about 100 gm of the food). But it is important to realize that potent levels of toxin can remain after viable cell numbers have declined naturally or after pasteurizing heat treatment. Indeed the toxins retain potency even after exposure to 100°C for half an hour.

### Effect on consumer
The incubation period before food poisoning symptoms appear can be 1 to 6 hours but is usually 2 to 4 hours. The symptoms are nausea,

cramps, diarrhoea and, in severe cases, prostration. The duration of the illness is usually 24 hours.

## Bacillus cereus

The first complete accounts of food poisoning due to *Bacillus cereus* related to incidents in Norway occurring between 1950 and 1955. Since then outbreaks have been reported from the Netherlands, Norway, Sweden, Hungary, USSR and the USA. *Bacillus cereus* was first definitely incriminated as a cause of food poisoning in Great Britain in 1971, in an incident involving the storage of cooked rice. Since that time *Bacillus cereus* incidents have been regularly reported. Although the organism is now firmly recognized as a cause of food poisoning, in the decade 1970–9, *Bacillus cereus* and staphylococci together were only responsible for 8 per cent of all outbreaks in Great Britain. The majority were caused by salmonellae and *Clostridium perfringens*.

### Morphology
*Bacillus cereus* is a gram positive, rod-shaped aerobic organism which forms heat resistant endospores and causes food poisoning through the production of toxins in the food in which it grows.

### Habitat
*Bacillus cereus* is commonly isolated from soil and from vegetation. It has been shown to be part of the normal flora of red and brown lentils, yellow and green split beans, black-eye beans, kidney beans, soya beans, pearl barley and chapatti flour, at levels of $1 \times 10^2$ to $6 \times 10^4$ colony forming units per gram, and is also found in rice and in spices. It has also been isolated from vegetable and meat soups, cooked meat, poultry, vegetables, puddings and sauces, cereals, grain and flour.

### Mode of infection of food
In foods which have been shown to have caused *Bacillus cereus* food poisoning, the organisms are most likely to have originated from that food as part of its natural flora. Most outbreaks of this type of food poisoning have been associated with cooked, reheated foods in which the spores were originally present in an ingredient. The spores survived cooking and subsequently germinated, and the vegetative cells produced then multiplied. *Bacillus cereus* is also responsible for a spoilage condition in cream known as bitty cream. It is notable that the foods incriminated in food poisoning outbreaks caused by *Bacillus cereus* are not reported to have changed either in flavour or appearance in spite of, sometimes, cell concentrations of $10^6$/gm or more. Such foods, of course, act as a source of organisms from which cross-contamination of other foods might arise.

### Growth requirements and environmental resistance
The vegetative cells of *Bacillus cereus* can grow aerobically in most food products of pH 6 to 7, in the temperature range 10 to 48°C (optimum 28 to 35°C), and during growth they produce toxins which are excreted into the food. The organisms can also grow in a wide range of oxygen tensions, but are inhibited in anaerobic conditions such as occur in cooking or cooked meats. Cooking usually destroys the vegetative cells, but exposure to temperatures just in excess of 70°C for 15 to 30 minutes will activate 50 per cent or more of the spores.

### Enterotoxin production
The toxic factors are produced during the growth of the vegetative cells and therefore accumulate in foods kept at suitable temperatures for a suitable length of time. Any cell density above $10^5$ cells/gm could be associated with significant enterotoxin concentration, and so should be considered as potentially dangerous. In special circumstances the enterotoxin produced by *Bacillus cereus* may occur preformed in the raw food. It has been shown, for example, that by soaking dried beans in water at 22°C, spore germination followed by growth to approximately $10^5$ organisms/gm can occur within a soaking period of 24 to 36 hours, and to $10^6$ organisms/gm within a soaking period of 48

hours. If the beans and the water in which the beans were soaked were used, the toxins would be present in the raw ingredients prior to cooking. At least two toxins appear to be produced – an emetic toxin producing vomiting and a diarrhoeal toxin – which differ not only in their effects but also in their susceptibility to heat. Temperatures below 100°C cause little to no diminution of the emetic toxin of *Bacillus cereus* because it is very heat resistant and can, in fact, withstand exposure at 126°C for 90 minutes. However, the diarrhoeal toxin is heat sensitive and its activity is markedly reduced at temperatures above 45°C, and exposure at 55°C for 20 minutes has been shown to eliminate its activity.

### Effect on consumer

The symptoms of *Bacillus cereus* food poisoning are predominantly nausea, vomiting and abdominal pain occurring around 12 hours after consuming the infected food. The symptoms can, however, occur much sooner. The illness is short, lasting around 12 hours. The minimum dosage thought to be necessary to produce symptoms has been put at approximately $10^7$ cells. Experimenters who have consumed cultures, taking between $4 \times 10^8$ and $2 \times 10^9$ organisms, have suffered acute symptoms within around 12 hours. Young children may be affected by lower toxin levels.

## Clostridium botulinum

Outbreaks of food poisoning due to *Clostridium botulinum* occasionally occur in the UK and in other parts of the world. The foods most frequently incriminated are raw and smoked fish products, and canned or home-bottled vegetables, meat and fish. In food canning, the industry either takes steps to ensure destruction of the vegetative cells and spores or prevent the formation of the toxin.

### Morphology

This organism is another member of the genus *Clostridium* and shares the same morphological characteristics, being a gram positive, spore-bearing anaerobic bacillus. At least seven strains of *Clostridium botulinum* have been identified, ranging from A to G according to the serologically distinct type of toxin produced. Most of the outbreaks recorded associated with man have been caused by types A, B, E and F.

### Habitat

Spores of *Clostridium botulinum* can usually be isolated in low numbers from most soil and mud samples. Some soils show relatively high counts, for example, as was demonstrated when the soil beneath a long disused cattle market and abattoir was examined, and shown to contain levels of four to five times the usual numbers. The organism thrives in the dead carcasses of animals and presumably reaches the soil in this way. Spores also occur at very low frequency in human and animal faeces. *Clostridium botulinum* type E, for example, has been isolated with some regularity from the intestines of fish and sea mammals.

### Mode of infection

The most likely manner in which food is infected with the organisms is by contamination with soil or faeces containing the spores, thus putting crops and vegetables at risk, as well as raw meat and fish.

### Growth requirements and environmental resistance

*Clostridium botulinum* is a strictly anaerobic organism capable of growth over a wide temperature range of at least 10 to 50°C, and, in some instances, over a wider range. Optimum growth occurs around 30°C. During growth toxin is produced which accumulates in the medium in which the organism is growing. Both growth and toxin production are affected by the pH and are generally considered to be completely inhibited below pH 4.5. In some specific instances pH values slightly below 4.5 may be tolerated. The organism is adversely affected by reductions in water activity, growth being progressively impeded with increasing sodium chloride concentration to complete inhibition at 10 per cent w/v, and similarly with sugar concentrations of 50 per cent w/v (that is, $a_w$ value in the region of 0.93).

The spores of *Clostridium botulinum* are heat resistant and able to withstand temperatures in the vicinity of 100°C for several hours. If the spores are given suitable conditions for germination, the vegetative cells emerge rapidly and produce toxin in the food. The toxins are more sensitive than the spores to the effects of heat and toxicity can be reduced by a factor of between 100 and 1000 within 30 minutes by heating at 80°C, or within minutes at 100°C. Variations in inactivation times obviously depend on the initial concentrations of toxin and the environmental conditions.

### Effects on consumer

The toxins of *Clostridium botulinum* are very potent, and affect the central and peripheral nervous systems. The incubation period is 12 to 36 hours and, if death occurs, it is usually within 24 hours and sometimes as long as 8 days. The symptoms include headaches, dizziness, tiredness, diarrhoea at first and, later, constipation. The central nervous system becomes affected and there is paralysis of throat muscles, speech becomes difficult and vision blurred. If an antitoxin is administered in time, life may be saved. It has been estimated that the toxin contained in only 2600 vegetative cells of *Clostridium botulinum* would be enough to kill a baby weighing about 7 kg. If this estimate is adjusted to allow a factor of 1000 to 1,000,000 for incomplete absorption of the toxin and weaknesses in the calculation of the number of toxin molecules per cell, perhaps $10^6$–$10^9$ cells (total) are sufficient to kill a baby; and $10^7$–$10^{10}$ cells to kill an adult weighing 70 kg.

## Campylobacter species

*Campylobacter* enteritis in man is a condition only recently recognized as being associated with the eating of infected food, although some diarrhoeal conditions in cattle and other animals have been known to have been caused by these vibrio-like organisms for many years.

### Effect on consumers

In man the illness starts as a 'flu-like' condition with fever, headache and general aches lasting from a few hours to 1 or 2 days. This is followed by a phase of profuse diarrhoea and abdominal cramps during which vomiting is largely absent. This diarrhoeic phase lasts 1 to 3 days. Recovery takes several days during which time the symptoms abate. Adults, children and babies are affected, the highest incidence being in infants.

### Morphology

Campylobacters are curved rods, similar to members of the Spirillaceae and rapidly motile.

### Growth

One group of Campylobacters shows a high optimum growth temperature (42–43°C), and it is this, the thermophilic group, which is responsible for acute enteritis in man and some animals.

### Habitat

Their normal habitat is the intestinal tract of many types of animal, including domestic dogs and, especially, birds and chickens. Campylobacters have been implicated in water borne and milk borne enteritis. One of the reasons for the fairly recent general recognition of Campylobacter (1977) as a cause of enteritis is that special methods for their detection are needed. They do not grow readily on normal laboratory media because, compared with other better known pathogens, they have unusual growth requirements. However, suitable methods for their detection and isolation have now been established.

Two outbreaks of Campylobacter enteritis were shown in 1978 owing to consumption of unpasteurized milk. In one case 63 people were affected and Campylobacters were isolated from the faeces of 38 of the 53 patients tested. These infections were thought to be due to thermophilic Campylobacter organisms, contamination to the milk occurring via the cows, by faecal infection or, possibly, by udder infections. Rectal swabs from 9 of the 85 cows tested were positive for the organism.

It has been shown that water can become contaminated, be a source of infection and the cause of an outbreak of enteritis.

A 'campylobacter red meat survey' conducted by the Public Health Laboratory Service (PHLS) in 1979 showed samples of raw red meat (beef, pork and lamb) to be contaminated and showed from where the infection could directly or indirectly have been contracted. Analysis of reports to the Communicable Disease Surveillance Centre carried out by the PHLS in 1977 showed infection was directly attributed to chicken and pork.

It is believed that only a low infective dose is necessary and that cell multiplication occurs in the intestine. The oral route of infection is the most likely but other routes of infection – through the skin for example – have not been ruled out.

### Destruction

Exposures to temperatures similar to those required to destroy salmonellae, are adequate to destroy this organism.

## Escherichia coli

This is an organism which characteristically lives in human and animal bowels and occurs in their faeces. It is therefore used as an indicator of the sewage pollution of water (see page 168). *E. coli* does not live very long (a matter of hours) in pure water; but in foods can survive and in some cases multiply. Many raw foods, especially those of animal origin, contain *E. coli* and therefore contamination originates from them and is exacerbated by cross-contamination through contact with infected surfaces, vessels and hands. The presence of coliforms in food is a sign of unhygienic food handling and the presence of *E. coli* a potential, but not definite, sign of faecal pollution. Apart from other pathogens which could well be present in such circumstances the *E. coli* themselves could give rise to abdominal upset with severe diarrhoea and possibly vomiting.

The condition in children, especially babies, can be serious or fatal and can be contracted through the use of dirty milk feeding bottles and teats. Proper bottle and feed sterilization procedures – both in hospitals and homes – markedly help to reduce the likelihood of this illness.

The invading organisms colonize and proliferate in the small intestine and cause an outpouring of fluid into the gut which shows itself as severe diarrhoea and dehydration.

The pathogenic strains produce several enterotoxins in the gut which are probably the cause of the symptoms. Young animals (piglets, lambs, calves and chicks) pick up the infection from the faecally contaminated environment in which they are born and spend their first days. Since it can be fatal it represents an environmental problem for stock breeders.

## Streptococci

*Streptococcus faecalis* and related species are mainly associated with the human intestines. Occasionally enterococci are reported as responsible for some food borne infections. For example, an incident occurred in 1975 in which 300 people contracted pharyngitis following a Thanksgiving Day dinner, which was possibly due to eating infected shrimp cocktail. Other outbreaks of food borne streptococcal infections have been reported due to eating infected cheese, cured ham, barbecued beef, Vienna sausage and turkey.

Streptococci are more heat resistant than coliforms or salmonellae, but are usually destroyed by accepted pasteurization processes or by cooking.

## Vibrio parahaemolyticus

This is a marine organism which can cause acute gastroenteritis. It is, in fact, the commonest cause of food poisoning in Japan – a country whose population consumes a lot of raw or lightly cooked sea food (oysters, clams, shrimp, prawns, crab, lobster, fish).

*Vibrio parahaemolyticus* is a gram negative motile vibrio which needs a saline environment in which to live. It can multiply (at 18–22°C optimally) in infected foods. The organism has been found in British coastal waters, although the low water temperatures discourage it, and occasional outbreaks of food poisoning associated with eating sea foods have been reported in Britain. It is probably only because fish and sea food form a relatively small proportion of our national diet that the number of incidents is small.

It is believed that sea products from temperate and warm seas are likely to be contaminated with *Vibrio parahaemolyticus*. It is sensitive to heat and effective reduction in numbers will be achieved by heating to pasteurization temperatures. Considerable care has to be taken in the processing of cooked crab and lobster meat to ensure that viable cells of the organism are not present, and subsequently preserved, in the period of deep freezing.

## Yersinia enterocolitica

Strains of *Yersinia enterocolitica* are associated with several types of human disease including septicaemia, eye and skin infections, and gastro-enteritis. The organisms are commonly isolated from the intestines of healthy cattle, other animals and human beings, and also from foods, such as raw milk, pasteurized milk and cream, raw meats, mussels, oysters, salad vegetables and water. Although not all strains of the organism are human pathogens, because they are widely isolated from foods, and can grow in refrigerated conditions at 0 to 5°C, their presence in foods is regarded with concern. They are heat sensitive and should be destroyed by correct pasteurization procedures. When organisms are found in heat-processed foods either underprocessing or recontamination (due to poor hygiene in production) must have occurred. Currently the relationship between the occurrence of the organisms in foods and the incidence of human gastro-enteritis caused by them is not known, but is being studied. Milk, for example, has sometimes been implicated in outbreaks of human gastro-enteritis but the one outbreak of Yersinia gastro-enteritis which occurred in the UK between 1970 and 1979 was probably due to the consumption of coleslaw salad contaminated by an infected food handler. It is probable that in the next few years the epidemiology of this organism will be more clearly understood.

## Mycotoxic fungi

In their growth many moulds and fungi may produce toxic waste products of metabolism. These mycotoxins diffuse into the surrounding substrate material, often well beyond the area visibly affected by mould growth, and remain there. There are increasing numbers of mycotoxins being identified, produced by a very wide variety of species and strains of moulds and fungi, particularly members of the genera Aspergillus, Fusarium and Penicillium.

In high doses, mycotoxins cause acute diseases, and in lower doses have been shown to cause carcinogenic, teratogenic and oestrogenic effects in experimental animals. Table 13 shows the known and probable effects of mycotoxins in human beings. Animals, particularly when they are intensively reared, may also be affected.

Moulds are capable of growth on a very wide variety of foods and animal feed stuffs. Surveys conducted by laboratories worldwide have shown the presence of spores of mycotoxic species in a comprehensive range of agricultural products and consumer foods (see Table 14). Problems with mould growth, and the probable production and accumulation of mycotoxins, are likely to arise when foods are stored. Growth will occur under specific conditions of moisture and temperature and will probably be accompanied by toxin production. However it has been shown in some cases that toxin production is progressively inhibited by increasingly adverse growth conditions, and toxin production is entirely inhibited before growth is.

Mycotoxins can be detected by several analytical techniques of which high pressure liquid chromatography is probably the most rapid. The toxins are destroyed by quite severe heat treatment, but survive mild processes. The best protection against mycotoxins, therefore, is to store vulnerable foods, especially those which

Table 13  *The known and probable effects of some mycotoxins*

| Mycotoxin | Effect |
| --- | --- |
| **Effect on human health** | |
| | *Confirmed conditions* |
| Ergot alkaloids (*Claviceps purpurea*) | Ergotism |
| Fusarial toxins (*Fusarium spp.*) | Septic angina |
| Psoralens | Contact dermatitis |
| Stachybotryotoxin (*Stachybotrys spp.*) | Dermal toxicity, leukopenia |
| | *Circumstantial evidence for conditions* |
| Aflatoxins (*Aspergillus spp.*) | Kidney dysfunction, carcinoma |
| | Liver damage and increased risk of carcinoma |
| Luteoskyrin | 'Yellow rice toxicity' |
| Citrinin and other mycotoxins | Nausea, vomiting, drowsiness, haemorrhage |
| **Effect on animal health** | |
| Aflatoxins (*Aspergillus spp.*) | Liver damage |
| Citrinin (*Penicillium spp.*) | Affects kidneys |
| Ergot alkaloids (*Claviceps purpurea*) | Affect blood system – leads to gangrene in the feet of cattle, sheep, poultry |
| Neurotoxins (e.g. cyclopiazonic acid) | Affect nerves – leads to uncoordinated movement in hind quarters of calves, piglets |
| Ochratoxins | Affect kidneys – structure and function |
| Trichothecin (*Fusarium spp.*) | Affects white blood cells – can lead to bowel haemorrhage |
| Zearalenone (*Fusarium species*) | Reduced fertility in dairy cows, laying hens |
| | Stillbirths |
| Other toxins | Alimentary disorders, for example |

Adapted from B. Jarvis, 'Mycotoxins in food', *Microbiology in Agriculture, Fisheries and Food*, eds. F. A. Skinner and J. G. Carr, pp. 251–64 (Academic Press 1976). A. Hacking and J. Harrison, 'Mycotoxins in animal feed', *Microbiology in Agriculture, Fisheries and Food*, pp. 243–50.

Table 14   *Foods liable to contain mycotoxins*

| Food | Mode of infection and growth characteristics |
| --- | --- |
| Cereals – rice, wheat, maize, oats, rye<br>Nuts – peanuts, pistachios, almonds<br>Oil seeds and meals<br>Ripe and stored fruits<br>Vegetables | Mould growth on the growing crop. In storage high relative humidity encourages mould growth |
| Meat, milk, dairy products | Accumulated residues due to the animals feeding on feeds containing mould or mycotoxin |
| Mould ripened cheeses and meats | Growth of mycotoxic species as well as the ripening species |
| Stored manufactured products such as cheeses, marzipan, bread | Growth of contaminants capable of producing mycotoxins |

*Source*: Jarvis, *op. cit.*

might not be subject to any heat treatment, by methods which discourage the growth of moulds.

# Chapter 5

# Food borne infections

There are many diseases spread by food which differ from microbial food poisoning in that the food, or water, acts merely as a means of transport for the organisms and not as a medium for growth. Small numbers of the organisms may be sufficient to cause infection, and examples can be found among the protozoa (toxoplasmosis), the bacteria (typhoid fever, paratyphoid fever, dysentery, tuberculosis, brucellosis), the viruses (Q fever, poliomyelitis, infectious hepatitis), and the parasitic worms (trichinosis).

**Bacteria**

Certain organisms of the genus Salmonella in addition to causing food poisoning also cause enteric fevers such as typhoid and paratyphoid fevers. These are true infections in the sense that the organisms invade the body tissues and become systemic. The incubation periods are longer than for food poisoning – up to three weeks in the case of typhoid fever – so the source of infection is often difficult to trace. These infections usually arise from the pollution of water, milk or other foods by sewage containing enteric organisms from human excretors. Paratyphoid fever may occur in two forms; one gives rise to enteric fever while the other produces symptoms of food poisoning similar to those caused by other food poisoning organisms of the salmonella group.

There are a number of foods which are capable of acting as vehicles of bacterial food borne diseases. Shellfish may act as a source of typhoid bacilli as they often grow in sewage-polluted river estuaries. Drinking water may become contaminated with sewage as a result of earthquakes, flooding and other similar disasters, or as a result of inadequate filtration and chlorination. Milk has long been known to act as a vehicle of food borne diseases. It may be contaminated by humans, from the cow via the udder and utensils or from water used in dairies. Typhoid and paratyphoid bacilli are capable of surviving and multiplying in milk at normal temperatures. Bovine tuberculosis used to be transmitted by milk, but now, with tuberculosis-free herds, this danger has largely been eliminated. Herds still suffer from brucellosis, but a national scheme to eradicate this disease from herds is progressing well. Recent research indicates that as the eradication scheme progresses so the number of human cases reported each year declines. The organisms can be transmitted through meat and through milk. Pasteurization of milk destroys the causative organism *Brucella abortus*, along with other pathogens that may be present. Cream (both fresh and imitation) and ice-cream have on occasions been incriminated as vehicles of enteric pathogens, especially of paratyphoid bacilli. Ice-cream is now compulsorily pasteurized and so this danger is eliminated, unless the ice-cream is recontaminated. Bakery products containing synthetic cream used to be common vehicles of paratyphoid organisms in Britain. Imported dried and frozen egg products, widely used in the bakery industry, were frequently contaminated with paratyphoid organisms and other salmonellae. The eggs were mainly used to make products which were subjected to heat which destroyed any salmonellae present. But it was found that on occasions paratyphoid bacilli in the raw egg mix were transferred to other food materials, such as synthetic cream, which were not subjected to heat. Frozen liquid egg is now pasteurized – a requirement under the Food and Drugs (Control of Food Premises) 1976 Act (See Appendix 1).

Occasionally, hitherto unsuspected foods become incriminated in outbreaks of food borne disease. It is now known that babies sometimes suffer from a condition known as infant botulism, in which the organisms have been picked up from food. The organisms of *Clostridium botulinum* proliferate in the infant's large intestine and produce toxin there. This disease has the symptoms of constipation followed by neurological signs, but fortunately, it is usually a mild condition and not often fatal. It was first recognized in the USA in 1976 where it is now estimated that about 250 cases per year occur. Following recognition of the condition, a wide variety of infant foods which had been fed to affected children were investigated, and they showed only one source of organisms – honey. Two further surveys of honey samples in the USA showed 10 per cent and 7.5 per cent of samples to contain *Clostridium botulinum*. It is believed that the honey merely acted as a vehicle for the spores of the organism. Isolated cases of infant botulism have also been reported in Australia and in the UK.

# Viruses

It has long been suspected that food can act as a vehicle of virus infection for diseases such as poliomyelitis and infectious hepatitis. This has now been proved in the case of infectious hepatitis through observing the accumulation and survival of the virus particles in shellfish. This has led to the realization that other virus particles might accumulate in the same way and be responsible for some incidents of food poisoning. Firm evidence for this theory is still lacking, although proof is increasing. But in the transmission of virus disease the food or water acts solely as a vector, and the viruses do not multiply.

## Viral hepatitis

The virus particles of classical viral hepatitis type A occur in human faeces (along with about a hundred other species of virus which include polio virus, herpes simplex virus and adenovirus). There have been a number of water borne epidemics of infectious hepatitis, often occurring during and after flooding. In Delhi, for example, in 1955–6, 36,000 cases occurred in a seven week period after the main water supply had been heavily contaminated with sewage owing to flooding.

Food borne outbreaks of viral hepatitis have occurred intermittently, involving raw milk, potato salad, orange juice, custard and roast pork – food probably directly infected from infected food handlers (see Chapter 6). Also, from time to time, cases associated with oysters and other shellfish occur, and have been found to be due to the growth of the shellfish in sewage polluted waters. In 1973 an incident occurred in the USA in which raw oysters were the source of infectious hepatitis in at least 283 people. Investigations suggested that the oysters accumulated and retained the viruses from the water. It seemed that in feeding, the shellfish filtered the water, rejecting unwanted materials yet retaining the virus particles in much higher concentrations than they exist in the water. The investigations also showed that although the oysters eliminated *E.coli* from themselves in the space of a day or two when they were held in fresh water purification beds, they retained the viruses for as long as six weeks after the last exposure. In this way 'cleaned' oysters were the source of infectious hepatitis virus particles.

Another notable outbreak of infectious hepatitis occurred in 1980, this time in the UK, where over 100 people contracted the condition after attending a banquet. Thorough investigation showed that fresh strawberries consumed at the banquet had been the vectors of the virus particles. The strawberries had been grown in open fields using only artificial fertilizers.But there had been no lavatories supplied for the labour force of casual pickers, some of whom must have been carriers, who, presumably, had relieved themselves in the fields. In this case the food had acted as a vector of the particles, but had not concentrated them in the way the oysters had. Type A hepatitis is spread by faecal carriers, especially by infected food handlers, by infected water supplies and by infected foods. The particles are able to withstand the normal levels of chlorination in drinking water (about 1.5 mg/kg available chlorine). Thus normal treatment of water, particularly recycled water, is insufficient to destroy them. Much stronger chlorination – in the region of 500–100 mg/kg available chlorine – is effective, as is boiling. But the particles survive in chilled and frozen conditions and probably also survive in conditions of dehydration.

### Food poisoning viruses

In about a quarter of all reported cases of food poisoning, the causative organisms are not identified. In cases where the incubation period is much longer and the condition more severe, with vomiting and diarrhoea, than is common with bacterial food poisoning, it is thought that viruses may be the causative agents. But because of the long period of incubation there is rarely any food left for microbiological analysis. Traditionally viruses are not looked for in food poisoning investigations because they are not included in the possible causes. The techniques for isolating viruses are more complicated and laborious than those for isolating bacteria. But now that viruses have been added to the probable causes of food borne gastroenteritis this may lead to the development of routine virological methods which are simple and practicable for use alongside bacteriological methods in the investigation of suspect foods.

Where viruses have been looked for in both food and in the faecal specimens of the patients involved, some cases have occurred where small, round, rather featureless viral particles – 'rotovirus' – have been found in the paired specimens, although this is not absolute proof of the viral cause of the enteritis. Viruses were shown to be present in cockles and in oysters

Table 15 *Percentages of enteritis patients' stools positive for rotavirus*

| | Out-breaks | Patients tested | Number positive for the rotavirus in faecal specimens |
|---|---|---|---|
| Cockles | 7 | 82 | 71 (87%) |
| Oysters | 2 | 8 | 7 (88%) |
| Foods not shell fish | 5 | 53 | 12 (23%) |

*Source*: PHLS Reports (1981).

which caused enteritis in 1977 and in 1979 respectively. Where no other agent has been found to be the cause of enteritis, viral particles have sometimes been found to be present both in the food and in the patients' stools (see Table 15).

So evidence is accumulating which points to the real possibility that foods, particularly shell fish, may be vectors of viruses which cause enteritis. 'Winter vomiting' is a condition for which no agent has been positively identified but which may well be caused by a food borne virus.

### Other viruses

Another virus disease that has been associated with food transmission is poliomyelitis. Raw milk, some unwashed fruits and vegetables and infected water have all from time to time been incriminated. It is also thought that there are probably many other virus diseases that could be food borne, but for which an association with food has not been established.

## Parasitic diseases

Trichinosis is a food borne illness which can be fatal, and is caused by a parasitic worm *Trichinella spiralis*. It is transmitted to humans through eating infected, undercooked meat (such as pork and horsemeat) in which the cysts survive. A survey reported in the *British Medical Journal*, *1*, 1047 (1979) summarizes the major outbreaks in Europe since 1860.

Toxoplasmosis is another food borne illness which is caused by ingestion of a protozoan parasitic organism *Toxoplasma gondii* in undercooked infected meat. It can also be transmitted through eating contaminated soil on, for example, unwashed vegetables.

There are other parasitic worms and protozoal infections which may also be transmitted by foods.

The best method of preventing any food borne disease – bacterial, viral, parasitic worm or protozoan – is by controlling the source of infection, and preventing the contamination of soil, food and water. The following precautions, however, help to prevent food borne diseases:

1   Pasteurization of milk, egg, cream.
2   Sedimentation, filtration and chlorination of water.
3   Efficient sewage removal, and its effective treatment
4   Education of the operatives in food hygiene.
5   Protection of raw foods – such as vegetables – from faecal contamination.

# Chapter 6

# Food pathogens: transfer, susceptible foods and preventive measures

Food may already contain food pathogens on arrival at the kitchen or may become contaminated with them during preparation as a result of cross-contamination from an infected source. Given sufficient time and a suitable temperature the food poisoning bacteria may increase to numbers capable of causing food poisoning symptoms, provided the food is a suitable substrate. Food borne disease organisms will survive in the food and probably cause illness after it is eaten.

This chapter deals with this chain of events and the preventive measures which can be taken to break the chain at one or several points. (See Figure 51).

## Transfer

On arrival at the kitchen or the point of preparation the food may already contain pathogenic organisms and may therefore serve as the means by which other foods, previously free from harmful organisms, become infected.

Chapter 4 mentioned that raw meat is potentially a source of salmonellae and of other pathogens. Surveys of minced meat taken from butcher's shops confirm this fact. In a survey of minced meats carried out in New Zealand in 1977, 55 samples taken from both butchers and supermarkets were analysed. The samples had total viable bacterial counts of between $1 \times 10^6$ and $1 \times 10^7$ organisms per gram. Staphylococci, *Clostridium perfringens*, *E. coli* and coliforms were found in the majority of samples. In fact 20 per cent of the

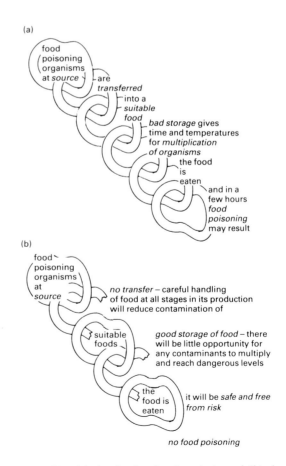

(a)

food poisoning organisms at *source* — are transferred — into a — suitable food — bad storage gives time and temperatures for *multiplication* of organisms — the food is — eaten — and in a few hours food poisoning may result

(b)

food poisoning organisms at *source* — no transfer – careful handling of food at all stages in its production will reduce contamination of

suitable foods — good storage of food – there will be little opportunity for any contaminants to multiply and reach dangerous levels

the food is eaten — it will be *safe and free from risk*

*no food poisoning*

Figure 51  *(a) the food poisoning chain and (b) the broken chain*

samples had more than 100 coagulase positive staphylococci per gram; 10 per cent of the samples showed *Clostridium perfringens* at 100 or more per gram, but *E. coli* and the coliforms occurred at more than 100 per gram in over 50 per cent of the samples.

Raw foods, such as meat and poultry, should therefore be handled with particular care as they are known reservoirs of all the food poisoning organisms and transmit pathogens, including viruses.

*The danger associated with raw meat is exemplified by an outbreak of infective hepatitis which occurred in Amsterdam in 1977. It was eventually traced to a single butcher's shop in which the butcher himself was suffering from the condition. A total of twenty people had been taken ill with infective hepatitis after having consumed meat or meat products from the shop. The connection between the illness and raw meat from the shop was emphasized by an incident involving a party of four girls who had bought meat there and had lunched together – three of the girls who had eaten steak tartare contracted hepatitis, while the fourth girl, who had abstained, remained well.*

Unpasteurized milk can act as a source of pathogens, including salmonellae, and Campylobacter species. Examples of illnesses arising from raw milk are given later in this chapter. Vegetables may be contaminated with soil organisms, such as spore-bearing clostridia, or with faecal organisms, such as salmonellae and viruses, if the vegetables have been fertilized with manure, either accidentally or purposefully (see page 91). Thus raw foods are an important source of pathogenic organisms, and only their proper handling will prevent the organisms from causing food poisoning (FP) or food borne disease (FBD) in the people who eventually consume the food. Transfer of the organisms from their sources to uninfected foods must be avoided.

Cross-contamination from raw to cooked food is particularly dangerous, as the latter will not necessarily be subjected to any further heat treatment.

In 1981 a restaurateur was prosecuted for keeping Black Forest gateau (a gateau made with cream and black cherries) next to raw cubed pork in the same refrigerator. If two such foods come into accidental contact, food poisoning organisms could be transferred from the raw pork to the gateau where they could well find the optimum environment in which to multiply. The public health authorities regarded this as a serious risk.

The chief ways in which cross-contamination is effected in the kitchen are by people and by equipment, surfaces and utensils. Traces of blood and meat on knives used in the preparation of raw foods can easily contaminate other foods if the knives are not sterilized between separate operations. In the mass production of food, cross-contamination is difficult to eliminate – the worktops, cutting machines, slicers, mincers and utensils may all become contaminated from food prepared on or in them and may pass the organisms on to other foods prepared with the same equipment later. Wooden surfaces such as chopping boards are particularly hazardous as they hold moisture, so providing an environment in which food poisoning organisms can survive and multiply and be present to infect other foods prepared on them subsequently.

Ignorance of the risks to food is no protection under the law in the UK. A case occurred in 1981 where the proprietor of a take-away food shop was fined a large sum of money because raw chickens had been delivered to the shop, had then been cooked and returned to the same cardboard boxes in which they had been delivered. The cooked chicken became contaminated with salmonellae which multiplied in the meat, and, as a result, twenty people who ate the chicken were taken ill with food poisoning.

Dish cloths, hand towels, savoy bags and other items made of fabric become damp in their daily use and are often impregnated with 'food solutions' which may contain food poisoning organisms. Experiments have demonstrated amply that many organisms survive in these cloths unless they are sterilized by boiling, or by soaking overnight in a disinfectant of adequate strength. Wherever possible, all items made of fabric should be replaced with disposable items, thus minimizing the opportunity of spreading pathogens.

Soil and dust may contain bacterial spores and salmonella organisms which may be brought into the kitchen in particles of soil present on vegetables and be spread by the operatives, surfaces or equipment to other commodities.

Up to 60 per cent of normal people are nasal and throat carriers of staphylococci and observations point to the fact that 15 to 20 per cent of nasal carriers also carry the organisms on their hands. In the course of their work these people may handle food and be using utensils and kitchen equipment which they could contaminate. The organisms are present in large numbers in skin infections such as boils, pimples, acne, infected cuts, barber's rash and so on, and these infections should therefore always be covered with waterproof dressings. Bad kitchen habits such as running the hands through the hair, picking the nose and fingering the mouth can transfer staphylococci to the hands and hence to food. Sampling foods during preparation is essential but could be the means of contaminating food if the spoon used is not changed or sterilized between each tasting. Smoking is prohibited in food preparation areas because of the risk of transferring infected saliva from the mouth to the hands and or any other surface with which the cigarette comes into contact, and hence to food. The hands are a common instrument of cross-contamination from raw to cooked food, a danger which can be minimized if cooked food is handled as little as possible. Washing the hands will not completely free them from staphylococci as the organisms are present in the pores of the skin.

The human bowel may harbour salmonellae, *Clostridium perfringens* and other disease-causing micro-organisms which are excreted from the body in faeces. A person may contaminate his hands from his own faeces after visiting the toilet because the toilet paper is porous, or the organisms may be derived from another person who had used the toilet previously, infecting the toilet seat, the chain, the toilet paper holder, the door handle or other places in the WC with which his hands had come into contact.

Less commonly, insects, rodents, and domestic pets may play a part in cross-contamination. Insects may contaminate food with their faeces and regurgitated food or they may act as vectors transferring organisms from an infected source to food. Rats may excrete salmonellae and, should their droppings come into contact with food, infection may result. Domestic pets, such as cats and dogs, are known excretors of salmonellae and may disseminate the organisms in the kitchen. (See Figure 52.)

## Susceptible foods

A food which will support growth of food poisoning organisms will be one which provides a nutrient environment. Consequently the foods which are most often associated with food poisoning are meat and poultry, fish and fish products, eggs and egg products, and milk and dairy products. (See Table 16.)

### Meat and poultry

#### Fresh raw meat and poultry

Raw meat and poultry may be the sources from which other foods, especially cooked foods, are contaminated via surfaces, operatives, and equipment. Surveys on imported boneless and carcass meat, for example, have shown up to 24 per cent of samples to be contaminated with salmonellae. Raw meat which is broken up in some manner tends to have more organisms per gram than the original carcass, as the contaminants are spread throughout the meat on to a greater number of exposed surfaces where they may grow. The process of boning and rolling joints may introduce contaminants into the central regions of the meat.

Table 16   *Types of food implicated in general and family outbreaks in England and Wales between 1976 and 1978*

**Presumed causal agent**

| Vehicle of infection | Salmonellae | Clostridium perfringens | Staphylococcus aureus | Bacillus cereus | Bacillus sp | Clostridium botulinum | Vibrio para-haemolyticus | All bacterial agents Number | All bacterial agents Per cent |
|---|---|---|---|---|---|---|---|---|---|
| **Meat** | | | | | | | | | |
| *Beef:* | | | | | | | | | |
| Stew and mince | 2 | 28 | – | – | 1 | – | – | 31 | |
| Shepherd's pies | – | 15 | 1 | – | – | – | – | 16 | |
| Reheated beef | 1 | 13 | – | – | – | – | – | 14 | |
| Steak pies | – | 13 | – | – | – | – | – | 13 | 96  23 |
| Cold beef | 1 | 8 | 1 | – | – | – | – | 10 | |
| Corned beef | – | 1 | 4 | – | – | – | – | 5 | |
| Other and unspecified | – | 7 | – | – | – | – | – | 7 | |
| *Pork and ham:* | | | | | | | | | |
| Cold pork | 4 | 8 | 2 | – | – | – | – | 14 | |
| Pork, veal and ham pies | 6 | 2 | 4 | 1 | – | – | – | 13 | |
| Reheated pork | – | 11 | – | – | – | – | – | 11 | |
| Roast pork | 2 | 4 | – | – | – | – | – | 6 | 57  14 |
| Ham and bacon | 1 | – | 5 | – | – | – | – | 6 | |
| Canned pork and ham | 1 | – | 1 | – | – | – | – | 2 | |
| Sausages and scotch eggs | 3 | – | – | – | – | – | – | 3 | |
| Other and unspecified | – | 2 | – | – | – | – | – | 2 | |
| *Mutton:* | | | | | | | | | |
| Reheated mutton | 1 | 6 | – | – | – | – | – | 7 | |
| Cold mutton | – | 5 | – | – | – | – | – | 5 | 14  3 |
| Other and unspecified | – | 2 | – | – | – | – | – | 2 | |
| Offal, tongue | 2 | 5 | 1 | – | – | – | – | 8 | 2 |
| Soup, gravy, dripping | 2 | 4 | – | – | – | – | – | 6 | 1 |

| Vehicle of infection | Salmonellae | Clostridium perfringens | Staphylococcus aureus | Bacillus cereus | Bacillus sp | Clostridium botulinum | Vibrio para-haemolyticus | All bacterial agents | |
|---|---|---|---|---|---|---|---|---|---|
| **Poultry** | | | | | | | | | |
| *Turkey:* | | | | | | | | | |
| Reheated turkey | 26 | 11 | — | — | — | — | — | 37 | |
| Cold turkey | 23 | 3 | 2 | — | — | — | — | 28 | 94 } 22 |
| Roast turkey | 8 | 2 | — | — | — | — | — | 10 | |
| Other and unspecified | 18 | 1 | — | — | — | — | — | 19 | |
| *Chicken:* | | | | | | | | | |
| Cold chicken | 18 | 13 | 8 | — | — | — | — | 39 | |
| Reheated chicken | 6 | 6 | — | — | — | — | — | 12 | |
| Roast and spit-roast | 5 | — | 1 | — | — | — | — | 6 | 82 } 19 |
| Chicken pies, vol-au-vent, casserole | 4 | 4 | 1 | — | — | — | — | 9 | |
| Other and unspecified | 9 | 3 | 4 | — | — | — | — | 16 | |
| **Other foods** | | | | | | | | | |
| Rice | — | — | — | 26 | — | — | — | 26 | |
| Milk, unpasteurized | 17 | — | — | — | — | — | — | 17 | |
| Sweets and cakes | 3 | 1 | 4 | 1 | — | — | — | 9 | |
| Seafood | 1 | — | 1 | 1 | — | 1 | 1 | 5 | 65 } 15 |
| Eggs, mayonnaise | — | 3 | — | — | — | — | — | 3 | |
| Vegetables | — | — | 1 | 1 | — | — | — | 3 | |
| Other foods | — | — | 1 | 1 | — | — | — | 2 | |
| **All foods** | 166 | 181 | 42 | 30 | 1 | 1 | 1 | 422 | 100 |

*Source*: E. Hepner, *op. cit. Public Health*, no. 94, pp. 337–49 (1980).

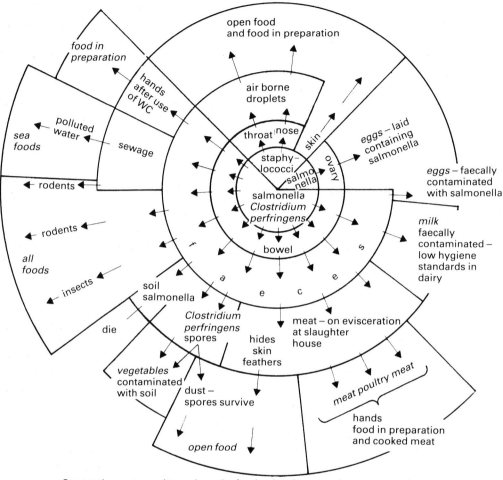

Start at the centre and trace how the food poisoning organisms are transferred directly or indirectly into raw and cooked foods

Figure 52 *Some of the paths of transfer of food poisoning organisms both directly and indirectly into food*

*Frozen raw meat, and poultry*

The freezing process will marginally decrease but not eliminate the organisms present and those that survive will be able to grow and multiply once the meat is thawed. Several recorded outbreaks of salmonellosis have been associated with cooking meat and poultry from the frozen state. The organisms were not destroyed because insufficient lethal heat was received at the centre of the food. It is advisable to thaw frozen meats thoroughly before cooking and to take care that the water which drips from them does not come into contact with other foods and contaminate them.

*Cooked meat and poultry*

From 1976 to 1978 43 per cent of the 422 general and family outbreaks in which a particular food was implicated were associated with meat, 41 per cent with poultry and the remaining 16 per cent with other foods. From these data it can be seen that meat and poultry are the foods most at risk. In the outbreaks associated with meat, beef was most frequently incriminated — it was associated with 47 per cent of the *Clostridium perfringens* outbreaks attributed to a particular food. Most of the beef was consumed in the form of stew, mince or meat pies, and many outbreaks were caused by lax catering practices – such as pre-cooking and then reheating the cooked meat after storing it at kitchen temperatures.

Meat may cause food poisoning when the organisms it contains are not destroyed by the cooking process. After cooking the organisms then have the opportunity to increase in number before the food is eaten. This occurs if the cooking temperature and time combined are insufficient to provide adequate lethal heat in the centre of the meat – for example, if a very slow cooking method is used or if the meat is very large in relation to the oven size. It could also occur if wholly or partly frozen meat is cooked, if a chicken or turkey is stuffed or if microwave cooking is improperly used.

*The dangers of inadequate thawing and slow cooling was well illustrated in a hospital outbreak involving 34 people where chicken vol-au-vent was the vehicle of infection. Both heat resistant and heat sensitive strains of* Clostridium perfringens *were isolated from the patients affected and from the chicken sauce mixture. The chickens had still been frozen on delivery to the kitchen and were thawed the following morning for 5–6 hours. They were boiled for 2–3 hours and then cooled in a larder for a further 3 hours. After cooling they were refrigerated at 4°C overnight. The following day the chickens were sliced and hot flour sauce was folded into the segments. They were served from heated trolleys at 12.15 p.m. having been ready for over an hour. It appeared that in this case the spores of* Clostridium perfringens *(both heat resistant and heat sensitive), originally present in the chicken carcasses, had survived the boiling, and germinated and multiplied during the subsequent slow cooling. Growth was probably further stimulated during the inadequate secondary heating of the chicken sauce mixture before serving. This outbreak could probably have been avoided if the chickens had been properly thawed before cooking, cooled rapidly after cooking and then reheated properly when the sauce was made.*

Although not infallible, a good indication of inadequate cooking of meat is if there is a flow of pink juices from any part of the meat when tested.

*An incident occurred in 1980 where the owner of a take-away shop was fined for selling cooked chicken 'unfit for human consumption'. The man pleaded guilty to selling chicken portions which contained undercooked blood pigments with a total viable bacteria count of not less than $1 \times 10^7$ organisms per gram. Such a bacterial count would not be possible in properly and freshly cooked meat.*

It has been suggested that the oven temperature for roasting should be between 190 and 204°C and 30 minutes cooking time should be allowed for every 0.5 kg of meat to achieve an adequate level of heat treatment.

*An outbreak of salmonellosis in Merseyside involved 160 people. The organism responsible was identified as* Salmonella virchow *and the source of infection was found to be cooked chicken from a shop selling it spit-roasted. The frozen carcasses were contaminated with* Salmonella virchow *on arrival and the outbreak was thought to be the result of inadequate thawing and cooking.*

Eight hours at room temperature has been suggested as adequate thawing time for small (less than 2 kg) frozen carcasses, or 24 hours in a refrigerator. Tests have shown that commercial spit-roasting of chickens often fails to satisfy the recommendations for adequate oven temperatures for safety because, as for oven roasting, time must be allowed for the spit to warm up.

Between 1976 and 1978, 70 per cent of the salmonella outbreaks in which the food source was established were associated with poultry. Of the outbreaks associated with turkey, 80 per cent were caused by salmonella and the major serotype was *Salmonella hadar*. The increase in the number of outbreaks associated with turkey is probably partly due to increased consumption, especially at large receptions such as weddings where catering facilities may be overstretched. The increase in outbreaks is certainly related to the emergence of certain salmonella serotypes, notably *Salmonella hadar*, in the turkey breeding flocks. Large outbreaks of food poisoning due to *Salmonella hadar* are becoming very common and the vehicle of infection is usually turkey, but occasionally it is chicken. It has become necessary to presume that poultry is contaminated with salmonellae and to take proper hygiene precautions to ensure that the cooked product is safe. Incidents of food poisoning arise from the use of big birds (often deep-frozen and inadequately thawed before roasting) for which the cooking heat may be insufficient to effectively sterilize the body cavity. Cooking is then followed by the slow cooling of the carcass as it is carved, which provides both favourable temperatures for bacterial growth and time for the organisms to radically increase in any meat left over after the meal. The importance of proper refrigerated storage for left-overs cannot be too strongly emphasized. The number of salmonella organisms eaten is important. If only a few organisms remain after cooking, immediate consumption may not cause food poisoning, but if the infected food remains warm for a while, the few organisms present may multiply and cause food poisoning.

*In 1980 an incident occurred in which, following a wedding reception, the remains of the turkey served at the reception were eaten by the waitresses. None of the guests were affected but the waitresses contracted salmonellosis. This implies that the salmonellae had grown in the cooked turkey in the warm conditions in which the meat remained during the reception.*

Staphylococcal food poisoning in meat products arises if the organisms reach the food after cooking and have the opportunity to multiply and produce toxin. The organisms grow well in salt beef, ham and poultry, and outbreaks occur from cured products as well. Handling of food seems to be the main way in which it becomes contaminated with staphylococci.

*In 1980, 98 people were affected with food poisoning after attending a buffet. About 180 staff and guests consumed a meal of fish, meat, salad dishes and creamed rice and most of the people suffered for 24 hours with acute stomach pains, diarrhoea and vomiting. A detailed investigation revealed that none of the members of the canteen carried the organism. The origin of the staphylococci was never discovered but the buffet food had been kept at room temperature for several hours before being eaten, and during that time it is believed that the original contaminants multiplied to produce concentrations of toxin which caused the food poisoning symptoms in the guests.*

*In the early 1970s, 50 guests attending a wedding had to receive hospital treatment after the reception. Laboratory tests showed that they were suffering from staphylococcal food poisoning. Staphylococcus aureus of the same type was isolated from the fingers of the caterer, from the chicken meat and jelly served, and from the vomit and faeces of some of the patients.*

*In another incident pressed beef proved to be the vehicle of infection. The meat was cooked, cut up and placed in tins and allowed to cool slowly overnight during warm summer weather. The pressed beef had been contaminated with staphylococci by the chef during the cutting up of the meat. As it was left unrefrigerated overnight the organisms multiplied and produced sufficient toxin to cause food poisoning when the food was eaten.*

*In an incident in 1980, brawn sold by a butcher caused acute enteritis in customers who bought it. The brawn was later shown to contain campylobacters which had presumably come from raw meat in the butcher's shop.*

If contaminated cooked meat is kept under conditions in which food poisoning organisms (of any type) are able to multiply, there is always a risk of food poisoning. There are many circumstances in which food may be left too long in warm conditions. These include keeping food in heated cabinets or in the kitchen overnight, warming of pies and previously deep-frozen stews and casseroles, putting food on plates long before it is served, displaying cooked food for sale in unrefrigerated counters and exposing food to the heat of the sun.

### Canned meats

A few types of canned meat, notably large

canned hams, are only subjected to temperatures akin to pasteurization and the heating is insufficient to sterilize the meat. Such cans should be clearly labelled 'Perishable – store in a cool place'. Most canned meats, however, are heat-treated sufficiently so that under normal conditions of storage they will not spoil, or harm the consumer. Once these cans are opened, however, they are subject to contamination and should be treated with the same precautions as ordinary cooked meat. The seams of cans are sometimes unsound and organisms enter (see page 173). Between 1976 and 1978 canned food was incriminated in only ten outbreaks. In five of these the vehicle of infection was imported corned beef. The causal organism in four outbreaks was *Staphylococcus aureus* and in the fifth incident, in which corned beef had been used as a pie filling, the causal organism was *Clostridium perfringens*. During the same period an outbreak of staphylococcal food poisoning was reported following consumption of canned chopped pork and an outbreak of salmonella food poisoning was reported following ingestion of luncheon meat.

A well publicized case occurred in 1979, in which imported cans of corned beef from Brazil were incriminated in causing food poisoning. The staphylococci, which caused the food poisoning were believed to have entered the cans at the factory in Sao Paulo and to have grown and produced toxin in the cans.

### Pies and sausages

Pork pies and similar products should be cooked thoroughly so that after cooking they have low bacterial counts. They may have gelatin added to them after being cooked. Gelatin is a very suitable medium for the growth of micro-organisms – a similar substance is, in fact, used as a medium in which to grow bacteria in the laboratory. Any organisms which are introduced into the pie at this stage may multiply profusely in the gelatin. If gelatin is boiled for 5 to 10 minutes before it is used in pies or for glazing and is then used immediately, or kept at temperatures above 63°C (see Appendix 1), it should be safe to use.

The unsafe use of gelatin was illustrated by an outbreak of food poisoning in the Midlands in 1976, involving at least 300 people. A *Salmonella senftenberg* infection was traced to large, manufactured pork pies which had become contaminated by jelly injected into the pies after cooking. If the gelatin had been properly used, and the pies chilled immediately after being filled with gelatin, it is possible that the outbreak would have been avoided.

The meat used in the manufacture of sausages is cut up before being mixed with meals and spices so that any surface contaminants will be distributed throughout the product. The organisms present in the centre may survive the cooking process and, in order to avoid this danger, sausages should be cooked so that they are browned on all sides and cooked in the middle.

### Fish and fish products

Only a small percentage of recorded outbreaks each year are caused by fish and other sea foods – approximately 1 per cent during the period 1976–8. Fish are not normally carriers of salmonellae, staphylococci or *Clostridium perfringens* because intestinal pathogens which attack humans and animals are unlikely to be present in creatures which have low body temperatures and which for the most part are caught a long way out to sea – away from sewage pollution. *Clostridium botulinum* type E has been isolated from fish in certain parts of the world, and outbreaks have been recorded resulting from the consumption of vacuum-packed smoked fish. The rather mild preservation undergone by the fish in the form of the smoking process, permitted the survival and growth in the package of *Clostridium botulinum* type E. All vacuum-packed fish should be refrigerated and the expiry date noted and adhered to.

In 1978 an incident occurred in which four people developed botulism and two died after eating canned salmon imported from Alaska. This was the first outbreak in this country since 1955 and the first incident in the UK associated with commercially canned fish. Type E toxin was proved to be present in the serum of all four patients and in the remains of the salmon; the same toxin is known to infect live salmon. The can was found to have a damaged seam which could have permitted the entry of the organisms and hygiene conditions at the cannery were found to be unsatisfactory.

Much less fish than meat is eaten in this country and most of that is eaten very soon after preparation and for these and other reasons previously mentioned, fish is less frequently incriminated as a vehicle of infection. Fish which has been cooked and made into something else, such as kedgeree, fish pie or fish cakes, may become contaminated with staphylococci from the hands of operatives, and as these dishes are usually cooked for a short period of time, some of the organisms may survive, multiply and produce enterotoxin.

Crab, lobster and shrimps may be contaminated with *Vibrio parahaemolyticus* before being cooked at the factory. Unless extreme care is taken there, the deep-frozen product sold from deep-freezes in UK shops may also contain viable vibrios. A case where this probably happened was notified in April 1980 where prawns from a single source gave rise to *Vibrio parahaemolyticus* food poisoning when they were eaten. The prawns had not been cooked after being thawed.

Proper cooking of crab, lobster and shrimps sterilizes them, but they are easily recontaminated afterwards if due care is not taken.

Shellfish may be eaten raw and do occasionally give rise to serious conditions such as typhoid fever and other intestinal illnesses. In 1976 and early 1977, about 800 people became ill after eating cockles. No bacterial pathogens were identified in either the patients or the cockles, but small virus particles were detected in specimens from patients.

As indicated in Chapter 5, shellfish, being filter feeders, may concentrate viruses from the sewage-polluted water in which they may have grown. Shellfish may also concentrate the dinoflagellate *Gonyaulax tamarensis* which can cause food poisoning.

Scombrotoxic food poisoning is associated with fish – an incident was recorded in 1978. Smoked mackerel is commonly associated with this form of food poisoning.

### Eggs and egg products

Raw egg is a medium in which salmonellae can thrive. Bulk liquid egg is being used increasingly in large bakeries, but since the introduction of the Liquid Egg (Pasteurization) Regulations, 1963 the danger of salmonella infection from this product has largely been eliminated. Even so, eggs of any type should be handled with caution to prevent cross-contamination. Bulk liquid egg may be sold frozen or in liquid form, in which case it should be refrigerated to below 5°C. Dried egg may be heat-treated before drying to reduce the number of salmonellae present.

When fresh eggs are being used, care must be taken to prevent cross-contamination from the shell or from the eggs themselves to other susceptible foods. This danger was illustrated by a family outbreak of food poisoning where mousse was the vehicle of infection, having been prepared with uncooked egg albumen. The organism responsible was *Salmonella typhimurium* and the eggs were home-produced. Salmonella of the same serotype were isolated from the bird's droppings, indicating that the hen was excreting the bacteria and had probably contaminated the outer surface of the eggs.

Desserts, sauces and mayonnaise containing uncooked egg should be handled with care, although in some instances sauces and mayonnaise may be protected by the acid pH of the vinegar they contain. Such foods, however, when they are *not* protected by their acidity, can provide an ideal medium for the growth of food poisoning organisms, for example, if the foods are contaminated during preparation, or if contaminated ingredients are used. In one outbreak where Hollandaise and Berenaise sauces were thought to be the vehicles of infection, *Staphyloccocus aureus* of the same type was isolated from both sauces, so it is possible that they were contaminated during preparation.

### Milk and dairy products

Milk and dairy products are nutritious foods in which food poisoning organisms can grow, if given the opportunity, and by which other organisms may be transmitted (see Table 17). Milk may be contaminated with staphylococci or coliforms from an infected udder, or from operatives; or with salmonellae or campylobacters from cows' faeces. All these organisms are readily destroyed in the pasteurization of milk but outbreaks of food poisoning do occur each year from raw untreated milk, or from products made from raw milk such as cream and cheese.

In 1979 a study was conducted in California on the association between raw milk and human *Salmonella dublin* infection. The results identified raw milk as the means by which people were infected. In this investigation 35 of the 44 cases involved had drunk raw milk produced by a very large herd of milking cows, though only a low percentage of the herd were shown to be carriers of *Salmonella dublin*.

An outbreak reported from Devon in the 1960s involving 50 people was caused by *Salmonella typhimurium* in untreated milk. It was thought that one apparently healthy cow had infected the milk through excreting the organism. The risk of further outbreaks was terminated by pasteurization of the milk, the exclusion of carriers and the isolation of the cow responsible.

Outbreaks of enteritis, due to a different cause, were also reported in 1979 and 1981 and were traced to the consumption of untreated milk. The 1981 incident occurred in Scotland, and affected about a quarter of the 600 pupils in one school. In 1979 an investigation into two outbreaks of enteritis in the north of England showed that all the patients had drunk untreated milk from local farms. The infection was found to be due to thermophilic campylobacters derived from a farm with a milking herd of 85 cows. The cows were machine milked and the milk then cooled to 5°C.

Campylobacter infections in cows have been known to occur for many years, but the organisms are not thought to be excreted with the milk. Faecal contamination of the milk is most likely to be the source of campylobacter to man. The incidence of campylobacter milk borne enteritis is increasing, and a considerable amount of investigation is being undertaken, particularly in relation to the infective dose, the ability of the organisms to survive in milk and the period of survival. At present it is thought that only a low number of organisms need to be ingested to cause enteritis. It is therefore very important that raw milk is handled hygienically and is, preferably, pasteurized.

Outbreaks of food poisoning from pasteurized milk are due to post-pasteurization contamination of the milk.

Fresh cream can be prepared from raw or pasteurized milk; it is usually pasteurized before packaging, but occasionally when in the bottle. Investigations into the bacteriological quality of fresh cream have shown that there is much post-pasteurization contamination. Properly pasteurized and hygienically packed cream will keep in good condition for 6 days at 5° C, but will not even keep for 2 days at 15° C. Correct storage, to inhibit the growth of the contained organisms, is therefore very important.

The composition of imitation cream varies, but it is usually made from dried milk, emulsified fats and a little sugar, and it is a product that is easily contaminated and liable to cause outbreaks of food poisoning. When used in products like trifles and cream cakes, the cake itself provides some nutrients for bacterial growth. Food poisoning due to imitation cream usually arises as a result of cross-contamination from infected ingredients, utensils, etc., in the kitchen. The pasteurization of manufactured imitation cream before canning now ensures that the cream is usually of good bacteriological quality, but cross-contamination from other infected products must be avoided in the kitchen as the cream can provide a good medium for growth.

Ice cream can easily be contaminated and it will support the growth of certain food poisoning organisms having in the past been responsible for staphylococcal food poisoning and also

Table 17  *Recorded outbreaks of communicable disease attributed to milk and dairy products in England and Wales between 1951 and 1980*

| Disease | Date | Cow's milk | | | | Cheese | | | Ice Cream | Total | |
| | | Raw | Pasteurized | Dried or Canned | Cream | Cow's | Sheep's | Goat's | | Outbreaks | Cases |
|---|---|---|---|---|---|---|---|---|---|---|---|
| Paratyphoid fever | 1951–60 | 2 | – | – | – | – | – | – | – | 2 | 156* |
| Salmonella food poisoning | 1951–60 | 13 | 1 | 6 | – | 1 | – | – | 4 | 25 | 1170 |
| | 1961–70 | 47 | – | 1 | 1 | 1 | – | – | – | 50 | 651 |
| | 1971–80 | 72 | – | – | – | – | – | 1 | – | 73 | 645* |
| Staphylococcal intoxication | 1951–60 | 17 | – | 3 | 6 | 11 | – | – | 1 | 38 | 2131** |
| | 1961–70 | 3 | – | – | – | 4 | – | – | – | 7 | 272 |
| | 1971–80 | – | – | – | – | 1 | – | – | – | 1 | 3 |
| Tuberculosis | 1951–60 | 1 | – | – | – | – | – | – | – | 1 | 3 |
| Brucellosis | 1961–70 | 1 | – | – | – | – | 1 | – | – | 2 | 15 |
| | 1971–80 | 1 | – | – | – | – | – | – | – | 1 | 2 |
| Q-fever | 1961–70 | 1 | – | – | – | – | – | – | – | 1 | 29 |
| Campylobacter infection | 1971–80 | 12 | 2 | – | – | – | – | – | – | 14 | 3983 |
| *Bacillus cereus* infection | 1971–80 | – | – | – | 1 | – | – | – | – | 1 | 2 |
| *E. coli* enteritis | 1971–80 | 1 | – | – | – | – | – | – | – | 1 | 2 |
| Outbreaks due to other organisms or an unknown cause | 1951–60 | 3 | 2 | 1 | 1 | 4 | – | – | 4 | 15 | 345 |
| | 1961–70 | – | – | – | – | 1 | – | – | – | 1 | 2 |
| **1951–80 total:** | | 174 | 5 | 11 | 9 | 23 | 1 | 1 | 9 | 233 | 9411**** |

* = 1 death

Adapted from N. S. Galbraith, P. Forbes and C. Clifford, *British Medical Journal*, no. 284, pp. 1761–5 (1982).

outbreaks of typhoid and paratyphoid fevers. The introduction of the Ice Cream (Heat Treatment) Regulations, 1959 (see Appendix 1) has resulted in the control of the factory production, pasteurization and sale of ice cream.

The use of pasteurized milk for making cheese has markedly reduced the risk of food poisoning from this source.

### Vegetables

Vegetables themselves are not able to support the growth of food poisoning organisms. However they often become contaminated with the organisms during their growth as a result of the practice of fertilizing fields of crops with animal and sometimes human manure. The foliage of vegetables may therefore be contaminated with organisms derived from the faecal material. This possibility has aroused much concern and has led to a number of surveys of fruits and vegetables. A Dutch survey was carried out in 1978 and it resulted in the isolation of salmonellae from 23 of the 103 samples. All samples were also tested for *E. coli* and *Streptococcus faecalis* as well as for salmonellae. A definite association was found between the number of these faecal indicators occurring per gram and the presence of salmonellae. The risk to tropical fruit and vegetables is higher because in the tropics the practice of manuring is more widespread. It is advisable to wash fruit and vegetables thoroughly in water of good microbiological quality. Care of fruit and vegetables after washing must also be observed. An incident occurred in 1980 where the juices from thawing frozen chickens poured into a sink which was used for washing salads and the juices infected the salads. The raw salad vegetables served in the restaurant caused about 40 cases of salmonellosis because the salmonellae on the vegetables had contaminated cooked meat, which was mixed with the salad vegetables and then eaten several hours after preparation.

Considerable care is taken in the production of commercially canned vegetables since the soil contaminating them may contain spores of *Clostridium botulinum* and the pH in the cans of vegetables is not usually low enough to prevent their growth. A massive outbreak of botulism, reported from the USA in 1979, was traced to a Mexican restaurant, and to a hot sauce made from home canned jalopino peppers. Home canning, not practised much in the UK and discouraged by the various food and health authorities, is a dangerous procedure because it cannot be monitored and controlled to ensure safety.

### Foods which are usually regarded as safe

It is perhaps dangerous to suggest that some foods are absolutely safe, but it is true to say that some are very much less likely to support the growth of food poisoning organisms than others.

Dry foods will not support the growth of food poisoning organisms but may contain them and any organisms present will be able to multiply and thrive once the food is rehydrated. Acid foods with a pH of below 4.5 are unlikely to support their growth, although some spoilage organisms are able to survive in acid conditions and some may even multiply. Foods like jam, syrup and honey have low $a_w$ values because they contain a high concentration of sugar and it is unlikely that they will support the growth of organisms which cause food poisoning. However 43 documented cases of infant botulism in California showed that 13 infants had ingested honey before the onset of the disease. The honey was eventually found to contain *Clostridium botulinum* spores (see page 69). The honey had acted only as a vehicle of transmission to the infants.

Fats and fatty foods are also unsuitable for the growth of food poisoning organisms but in some instances they have been known to survive in these foods and could cause food poisoning if they were transferred to a food more suitable for their growth.

Good hygienic practices rely on the supply of good quality water. In the UK the quality can usually be counted on at the point of use. However there are occasions when the water supply becomes contaminated, and water borne disease occurs. Such an incident occurred in

1980 in the Leeds area. Between two and three thousand people suffered gastroenteritis symptoms which were thought to be attributable to the use of water accidentally infected with sewage. Such contamination would not necessarily be visibly evident. In this incident the quality of the supply was rapidly restored to drinkable quality.

## Preventive measures

There are a large number of precautions which, if observed, would produce a reduction in the incidence of food poisoning in this country. Improved hygiene at all stages of meat production – on the farm, during transport, in the lairages prior to slaughter, in the slaughter house and in food premises – would help to reduce the problem.

This section deals with the prevention of food poisoning in food preparation areas. It should be assumed by food operatives that the raw food ingredients are probably contaminated on arrival at the kitchen and should be treated accordingly.

### Care of food

The foods to be treated with particular care on arrival at the kitchen are raw meat and poultry as these may act as a source of organisms. If they are not to be cooked immediately they should be placed in refrigerated storage.

### Refrigeration and cooling

Safe refrigeration requires the unit to operate at a temperature below +5°C, preferably between +1 and +4°C. It is wise to check regularly the internal temperature of a refrigerator because experience shows that the reading on the indicator thermometer is often inaccurate. A refrigerator is not usually designed to reduce the temperature of a warm food, but will maintain foods at its running temperature. Therefore if foods at temperatures above the running temperature of the refrigerator are transferred to the refrigerator, the effect will be to put a load on the cooling capacity which may result in a temperature rise in the refrigerator lasting

anything from a few minutes to several hours. The practice of placing hot or warm foods in a refrigerator can also cause condensation and dripping. Also, frequent opening of the cabinet puts a considerable strain on the cooling system and the running temperature will only then be maintained with difficulty.

Consideration should be given when planning catering establishments to the provision of rooms designed especially to cool large amounts of food quickly, in addition to the traditional type of cold room. Heat is lost rapidly at first from a hot food, and then more slowly as it passes through the dangerous temperature zone when food poisoning organisms are able to multiply (Figure 50). The time spent in this zone should be as short as possible, and the cooled food should then be refrigerated. Food poisoning organisms do not usually multiply at below 5°C but *Clostridium botulinum* type E has been reported to grow and produce toxin at a temperature as low as 3.4°C.

Cooked and raw foods should always be kept apart, ideally in separate refrigerators. Cooked meats and fish, gravy, sauces, cold desserts and dishes containing eggs or cream should be kept refrigerated.

### Thawing

Frozen foods on arrival at the kitchen should be stored in a deep-freeze with an operating temperature of −18°C. On use, frozen raw meats and poultry should be thawed either at room temperture or in the refrigerator. The time allowed for this should be adequate to ensure that no part of the meat is still frozen when cooked, but the meat should not be left too long, otherwise deteriorative processes will set in. Large turkeys, for example, used for banquets and similar occasions, may weight up to 16 kg, and may need two days defrosting in a cool room. For small birds and joints (up to 2 kg) 8 hours at room temperature or 24 hours in the refrigerator should be long enough.

### Heating and cooking food

When food is cooked it is subjected to heat which alters it both chemically and physically, and at the same time should destroy most of the bacteria it contains. Vegetative cells and the less heat resistant spores in the central regions of foods will only be destroyed if the heat penetrates adequately into these regions. In order to achieve this, several steps have to be taken. Raw foods should, as indicated above, be thawed properly all the way through.

The cooking methods which are considered to be the most safe are pressure cooking, roasting, grilling and frying. The less safe methods include braising and boiling. Meat should be cooked in small portions (less than 2.5 kg) and in the case of poultry the body cavity should not be stuffed, because this adversely affects sterilization of the body cavity.

Recently developed domestic methods of cooking foods need to be used with caution and instructions given in the manufacturer's booklets should be closely followed. Slow cookers should only be filled up to the maximum recommended capacity at which they are capable of cooking safely and pasteurizing the food. If a slow cooker is overfilled it is possible that the food might not be adequately cooked. Microwave cooking, which is used increasingly in catering, should be used with caution when cooking raw foods because the heat distribution in the food tends to be uneven.

Mention must also be made of another development in catering. A system such as 'cook-chill' involves the bulk cooking of food (often by microwave), followed by fast chilling, temporary storage at low temperature, and subsequent reheating when needed. It is obviously a system which requires very good temperature and hygiene control. As a result there have been many examinations of the microbiological and other aspects of this system (for example, *J. Microwave Power* (1978), no. 13(1), 87–93) and assessments have been carried out by users. The borough of Middlesborough introduced the system into the school meals service in the late 1970s and it reports that a satisfactory system is possible.

If all cooked foods were eaten immediately after cooking there would be a marked drop in the incidence of food poisoning. It is important

that meat if it cannot be served immediately is cooled as rapidly as possible and refrigerated at below 5°C. This is an easier procedure with smaller joints. The critical time is the period after the food has been cooked and before it is eaten. The nature of the food storage conditions determine whether any contaminants present will multiply. In a suitable food, which is kept at a temperature of between 5 and 63°C, it is theoretically possible for growth and multiplication of food poisoning organisms to take place. Under optimum conditions certain species can reproduce approximately every 10 to 12 minutes. Hot foods cooling at room temperature, or in a cool room, should not remain within this temperature range (5 to 63°C) for longer than 1.5 hours, and during this time should be protected from contamination by insects, rodents, domestic pets or raw foods. Raw and cooked foods should be kept entirely separate, hands should be washed, and utensils and surfaces sterilized after dealing with either type

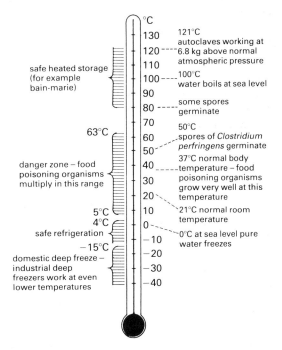

Figure 53  *The relationship between temperature and the growth of food poisoning organisms*

of food. No surface used previously to prepare raw foods should have hot uncovered cooked food placed upon it without being thoroughly cleaned and sterilized first.

*Reheating food*

Food should never be warmed up. Already cooked food should be recooked – that is, heated up to 100°C, especially if it is to be incorporated into some other dish. Only by heating to 100°C will any food poisoning organisms which might be present be destroyed. The warming of food will only encourage the multiplication of bacterial contaminants (see Figure 53).

*Care of person*

Good personal hygiene will reduce the likelihood of food contamination, and should be applied at all stages. Hands should be washed before and after handling food, as they can serve as a means of cross-contamination. Hands should be scrupulously clean and frequently washed with antiseptic soap, and nails should be scrubbed. Nail brushes, however, can act as a means of cross-contamination from one person to another and should be made of some material which can be easily sterilized. However it is dangerous to assume that washed hands are safe, for *Staphylococcus aureus* is not necessarily removed by washing and skin surface bacterial counts may even increase after washing because the organisms in the skin pores rise to the surface. Hands must be kept in good condition so that cracking and chapping do not occur, as these lead to skin infections. The use of hand creams helps to minimize this problem. Hands should be dried thoroughly, preferably by the use of disposable paper towels, for which a disposal bin should be provided, or by a foot-operated hot air drier.

Public authorities occasionally carry out surveys to assess the effectiveness of hand-washing techniques in catering premises. The borough of Middlesborough did so in 1980 in co-operation with the Public Health Laboratory Service. A survey of twenty premises was carried out in which finger rinse samples were obtained from

food handlers together with swabs from soaps, towels and nail brushes. Small numbers of faecal organisms were found on the hands of the staff, but this was considered to be only one of the methods by which organisms were conveyed to food – soaps, towels, etc., being responsible. Where soap dispensers containing a bactericidal agent were in use, very few organisms were found, but where bar soap and dispensers without a bacterial agent were in use, a higher incidence of organisms was found.

Personal habits should be good – no touching of the hair, nose or mouth during food preparation. Hair should be kept covered with a protective cap and clean overalls should be worn. Minor cuts and abrasions and spots and pimples on the face, hands and arms should be covered with waterproof dressings, or, where appropriate, a finger stall. This not only protects the food from infection but also protects the individual from organisms derived from the food. Smoking or taking snuff in a food preparation or service area is forbidden. No rings or other jewellery should be worn because they harbour dirt and bacteria and there is also the possibility that they may fall into the food. When tasting food a clean spoon should be used on each occasion. Wherever possible, implements should be used to handle food, minimizing the possibility of transferring organisms from the hands to food.

Any serious skin infection should be reported to a doctor and the person affected should not prepare or in any way come into contact with open food while in that condition. Similarly any diarrhoea-like condition should be reported to the supervisor and the sick person removed from work. The wisdom of keeping infected persons away from food preparation is exemplified by an incident, which was reported from the borough of Gosport in 1978, involving a 4 month outbreak of bacillary dysentery in which 122 cases were notified. The source was traced to an area and to a particular family where the mother was engaged in handling food for sale. The mother was a symptomless excretor of *Shigella sonnei* and her stools were positive for the organism. A second food handler became infected and both women were excluded from work by their doctor.

Another incident occurred in May and June 1977 in Kingston-upon-Hull. It involved an outbreak of salmonellosis due to *Salmonella agona*. The source was traced to *Salmonella agona* infected pork dripping which was supplied to a chain of retail shops by a cooked meat supplier, and it caused at least 38 cases of food poisoning. In the thorough investigations carried out at the factory and at the suppliers, 49 food handlers were shown to be symptomless excretors of *Salmonella agona*. These 49 people were excluded from work until they were cleared of infection, as were nine other infected people in high risk employment (nurses, teachers, food workers).

Many countries require that food workers undergo pre-employment medical screening, and some countries carry out statutory routine examination of the stools of food handlers. A few countries including the UK have no such legal requirements. There is much international debate as to the value of such routine screening and the contribution it makes to the prevention of food borne disease. While food handlers are often the victims rather than the cause of food infection, larger food concerns – factories, retail chains of shops and caterers – are increasingly holding the view that screening employees on a regular basis, particularly those returning to work after illness, does have a value in preventing symptomless excretors from being the cause of food borne disease.

### Care of equipment, utensils, surfaces

Equipment, such as knives, beaters and basins, and machines, such as slicers, mixers and conveyor belts, used to deal with raw foods should be cleaned and sterilized before being used for any other foods, particularly cooked food or products which are eaten uncooked, such as cream. Raw eggs, for example, may be contaminated with salmonellae which would be destroyed during cooking, but if the utensils used in dealing with the eggs are not sterilized after use, the organisms may be passed to other commodities which are prepared in them later. The methods of cleaning kitchen equipment, surfaces and utensils are discussed in Chapter 10.

The Department of Health and Social Security has prepared a useful ten-point code for food trade workers which is as follows:

---

*Ten-point code for food trade workers*

1  Wash your hands always before touching food and always after using the WC.
2  Tell your supervisor at once of any skin, nose, throat or bowel trouble.
3  Cover cuts and sores with waterproof dressings.
4  Wear clean clothing and be clean.
5  Remember, smoking in a food room is illegal and dangerous. Never cough or sneeze over food.
6  Clean as you go in food rooms.
7  Keep food clean, covered and either cool or piping hot.
8  Keep your hands off food as far as possible. Keep food utensils clean.
9  Keep the lid on the dustbin.
10 Remember, the law requires clean, fully equipped, well lit and airy conditions for food preparation

---

(Prepared by the Department of Health and Social Security, with the agreement of the Food Hygiene Advisory Council. Published by HMSO.)

# Chapter 7

# Principles of microbial spoilage

Much of the food which we eat is imported from abroad – such as meat from Australia and New Zealand, fruit from California and wheat from Canada – so that many of the foods consumed by the British have to be harvested in the country of origin, stored, transported long distances, processed and sold before they are consumed. These procedures take time and as foods are not stable they can deteriorate rapidly. When food is no longer attractive or safe to eat it is said to be *spoiled*.

## Causes of spoilage

The main causes of food spoilage are:

1  Physical damage in transporting, storage, etc., resulting in changes in texture such as bruising.
2  Insect, rodent or other animal activity.
3  Chemical breakdown or chemical contamination resulting in deterioration in quality.
4  Autolytic enzymes which catalyze reactions within the food resulting in texture breakdown and the food becoming soft and pulpy. It may also be more vulnerable to attack by insects, rodents and micro-organisms.
5  Micro-organisms, whose entry into the food is aided by 1, 2, 3 and 4 above, which grow and change the texture, colour, taste, smell and quality of the food.

Spoilage caused by micro-organisms is recognized by changes in foods which are often given common names – such as 'slime', 'rots' and 'putrefaction'. The main features of microbial spoilage are that the texture of the food degenerates and it gradually becomes soft and sticky and eventually fluid. These changes are often accompanied by odours which become more marked as time passes – some odours are very distinct, of which the 'sulphur stinker' spoilage of canned foods is an extreme example. Spoiled food can also change in colour, although this is dependant on the type of organism present.

The characteristics of these micro-organisms which cause them to spoil foods are those which also make them beneficial in the normal decay of organic material in the soil and in water. They break down the organic components of foods for their own use and in so doing convert them to simpler compounds.

## Types of spoilage

The main types of spoilage are:

### Mouldiness and 'whiskers'
Moulds, being aerobic, grow mainly on the outside surfaces of the affected foods, initially as small separate colonies – 'spots' which may later merge. Foods become sticky, 'whiskery' and locally coloured.

### Rots
A general word used to refer to spoilage of fruit, vegetables, eggs and other foods, for example, black rot of eggs, watery soft rots of fruit and vegetables.

### Sliminess

Growth of bacteria on moist surfaces of vegetables, meat, fish, etc., may cause *taints* and *odours* and can result in such deterioration of the food that it degenerates into slime. Pigmentation may occur at the same time.

### Colour change

Many microbes produce brightly coloured colonies or pigments which give colour to the spoiling food, for example *Serratia marcescens* – red, *Sarcina lutea* – yellow, *Pseudomonas fluorescens* – green with fluorescence, *Aspergillus niger* – black, *Penicillium* species – green.

### Ropiness

Rope is the formation of a viscous sticky material closely allied to slime and caused by a wide variety of organisms such as *Leuconostoc mesenteroides, Leuconostoc dextranicum, Bacillus subtilis, Lactobacillus plantarum* and others. In some foods, especially high sugar foods, the rope organisms produce copious capsules, and as the number of cells increases, the rope appears. Rope is also caused by microbial hydrolysis of starch and protein to produce glutinous non-capsular materials. Rope can affect soft drinks, wine, pickling brine, vinegar, milk and bread.

### Fermentative spoilage

Many types of organisms, especially yeasts, aerobic and anaerobic sporing bacteria, and lactobacilli are able to ferment carbohydrates. Yeasts usually convert sugars into alcohols and carbon dioxide; 'homofermentative' lactic acid bacteria convert sugars into lactic acid, while the heterofermentative bacteria produce several acids, such as butyric and propionic acid, in addition to lactic acid, and the gases carbon dioxide and hydrogen. *Bone taint* refers to fermentative spoilage which arises close to the bone in meat. *Flat sours* occur in canned foods in non-gas producing fermentative spoilage. *Blown cans* occur as a result of gas producing fermentation in which such copious quantities of gas are evolved that the pressure within the can distorts the sides and ends of the can and it may eventually blow (see Figure 54). Fermentative spoilage may occur in foods which are produced by fermentation – 'wild' organisms flourishing to the detriment of the product. This can be a problem in beer manufacture, for example.

### Putrefaction

The anaerobic decomposition of proteins into peptides or amino acids causes the production of foul odours in the food due to hydrogen sulphide, ammonia, methyl and ethyl sulphides, amines and other strong smelling products. Foods which are likely to deteriorate in this way are those which have been poorly processed and packed to provide anaerobic conditions – for example improperly processed canned meat and vegetables.

### Aerobic hydrolysis

Aerobic hydrolysis of proteins leads to the development of bitter flavours in foods – which are not necessarily unpleasant and sometimes enhance the flavour.

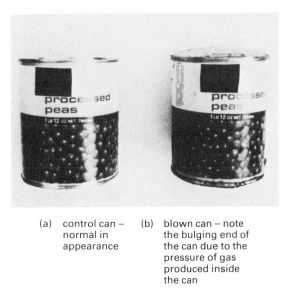

(a) control can – normal in appearance   (b) blown can – note the bulging end of the can due to the pressure of gas produced inside the can

Figure 54 *Canned food spoilage due to fermentative changes*

## The organisms

The way that spoilage develops in a food depends on the types of organisms present and whether the food, under its existing conditions of storage, can support the growth of any or all of them.

### Microbial load

The microbial load – the numbers and species of organisms which a food carries – is initially determined by the food type and its origins. Later it will be altered by the handling and processing to which the food is subjected.

Raw foods carry their own characteristic flora. Soil crops carry on their surfaces organisms which are saprophytic or parasitic, as well as soil organisms; meat carries organisms derived from the animal bowel, skin and fur; fish, organisms derived from the fish skin, intestine and from the water in which it lived; milk, organisms from the udder; etc.

Subsequent treatment of the food will either reduce or increase the total load. Procedures such as the removal of soil from root vegetables, peeling fruit or vegetables, washing foods and heat treatments, such as pasteurization, cooking or canning, tend to reduce the microbial load. Storage of food under warm conditions – such as grain in inadequately aerated silos, or fish or other food in a warm shop window – tends to increase the microbial load. Table 18 lists some treatments of food and their probable effect on the microbial load of the food.

Table 18 *Some treatments of foods and their probable effect on the microbial load*

| Food type | Treatment | Probable effect on microbial load |
|---|---|---|
| Grain | Harvesting | Adds organisms in dust to the natural flora |
| | Milling – removes outer layers | Reduces flora |
| Flour | Storage under dry conditions | Dormant organisms gradually reduce in numbers |
| | Storage under damp conditions | Moulds and yeasts very likely to proliferate |
| Milk | Kept warm | Numbers of bacteria increase |
| | Chilled immediately after milking | Numbers of bacteria increase slowly. Psychrotrophic organisms increase |
| | Pasteurized | Numbers of organisms reduced – heat sensitive organisms killed |
| | Sterilized | Cells destroyed – no growth |
| Bulk egg | Pasteurized, chilled | Very slow growth of remaining organisms |
| | Pasteurized, frozen | No growth while frozen |
| Vegetables | Harvesting | Adds organisms from dust and soil |
| | Washing | Removes some surface organisms |
| | Washing, followed by storage wet or damp | Organisms multiply leading to rapid spoilage |
| Fruit | Bruising | Aids penetration of organisms – leads to rapid spoilage |
| | Storage wet | Rapid increase in numbers of organisms |
| Meat | Warm storage | Rapid increase in numbers of organisms |
| | Prolonged chilling | Increase in numbers of organisms especially of psychrotrophs |
| | Frozen storage | No increase in numbers of organisms while frozen |
| Fish | Warm storage | Very rapid increase in numbers of organisms |
| | Wet cool storage | Increase in numbers of organisms |

### Inter-relationships between organisms

If a food contains a single species of organism, or only very closely related species, then their growth will not be in competition with any others and the growth rate will be determined by the environmental conditions. If spoilage occurs it will probably be very characteristic, such as the flat souring of canned foods by *Bacillus* species.

In a mixed population of organisms the different species affect each other's growth in several ways.

The rate at which an organism is able to multiply in a food determines whether it will achieve dominance, the fastest growing organisms having the greatest opportunity. When bacteria, yeasts and moulds are present in a food which is capable of supporting the growth of all three it is most likely that the bacteria will become dominant first. Mould or yeast spoilage may occur at a later stage if the conditions in the food at that time permit.

The waste products that the dominant organisms produce may either stimulate or inhibit the growth of other organisms present. For example, some moulds of the Penicillium species may produce antibiotics in their growth which are inhibitory to other organisms; some bacteria may produce acids which favour the growth of acidophiles.

Sequential spoilage occurs when the initial wave of growth due to one or several species of organism dies down due to factors such as overcrowding, depletion of food supply and build up of waste products to toxic levels (see page 36). The conditions now existing may favour the rapid growth of a second group of organisms whose growth up to this point has been repressed. In a similar way to the first wave, the second wave may later die down to be replaced by a third and also perhaps a fourth wave of growth. The spoilage of the food which results from these population changes may be distinctive – as when mould follows bacterial growth.

The conditions of storage and the treatment of a food affect which categories of organisms can become dominant, for example:

1  Pasteurization destroys heat sensitive bacteria, yeasts and moulds, and leaves heat resistant spoilage organisms.
2  Storage of food under chilled conditions may discourage mesophiles but allow psychrophiles to grow unchecked.
3  Vacuum-packed food can spoil anaerobically whereas packed aerobically it would spoil in a different way.

### Moulds in spoilage

Mould growth is initiated when a ripe spore is able to germinate and start mycelium growth. The affected food becomes coloured, musty, softer and sticky or slimy. Because moulds are aerobic, spoilage generally begins at the surface, although the mycelium later penetrates deep into the food. As well as spoiling the more perishable foods, moulds are often associated with the spoilage of 'dry' foods especially those stored under damp conditions and those foods containing high concentrations of sugar or salt.

### Moulds important in food spoilage

1  **Non-septate moulds**–reproduce by asexual and sexual spores.
   (a) *Genus Rhizopus* Widespread. Fluffy, luxuriant mycelium. 'Pin head' sporangia which become dark as they ripen. Spoilage: 'bread mould', soft rots in fruit and vegetables, spoil chilled meat. See Figure 23.
   (b) *Genus Mucor* Widespread. Approximately 150 species. Hyphae pale; sporangia become greyish as they ripen. Spoilage: wide range of foods affected. See Figure 27.
   (c) *Genus Thamnidium* Not common. Greyish brown sporangia. Psychrophilic. Spoilage: chilled meat. See Figure 55.
2  **Septate moulds** – usually reproduce by asexual spores only.
   (a) *Genus Aspergillus* Widespread. Compact colonies – white, buff, green, black. Bear conidia in 'globose' heads. Two important groups: *Aspergillus glaucus* group  grey, green. Grow well

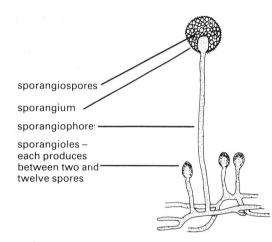

sporangiospores

sporangium

sporangiophore

sporangioles –
each produces
between two and
twelve spores

Figure 55   *Genus* Thamnidium – *non-septate*

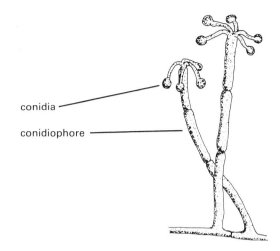

conidia

conidiophore

Figure 56   *Genus* Trichothecum – *septate*

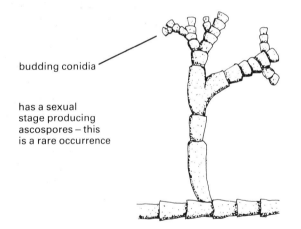

budding conidia

has a sexual
stage producing
ascospores – this
is a rare occurrence

Figure 57   *Genus* Monilia – *septate*

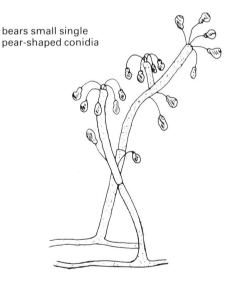

bears small single
pear-shaped conidia

Figure 58   *Genus* Sporotrichum – *septate*

in a low $a_w$. Spoilage: dried foods and those preserved in sugar and in salt. Optimal temperature range for growth 15–20°C.

*Aspergillus niger* group Black conidia. Spoilage: bread, black rots of fruit and vegetables. Optimal temperature for growth 30°C. See Figures 19 and 20.

(b)   *Genus Penicillium*   Widespread. Compact velvety grey-green, or white colonies. Bear conidia in 'brushes'. Spoilage: soft rots in citrus fruits, 'blue rot'; greenish patches on stored meat, yellow or green spots in eggs, greenish spoilage of cheddar and other cheese, bread, and so on. Optimal temperature

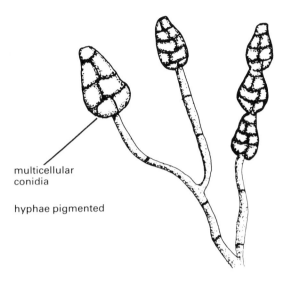

multicellular
conidia

hyphae pigmented

Figure 59   *Genus* Alternaria – *septate*

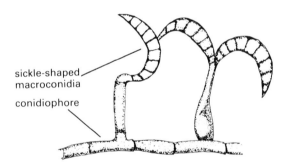

sickle-shaped
macroconidia

conidiophore

Figure 60   *Genus* Fusarium – *septate*

for growth 20–25°C. See Figures 21 and 22.

(c) *Genus Trichothecum. Trichothecum roseum* causes spoilage of stored moist fruits. See Figure 56.

(d) *Genus Geotrichum* Commonly found in dairy produce; compact felt-like colonies – white, yellow, red or orange. Hyphae break to form arthrospores. Spoilage: dairy produce – yoghurt, cheese, bread, stored citrus fruits and chilled meat. See Figure 26.

(e) *Genus Monilia* Associated with the spoilage of bread – pink loose textured growth requiring moist conditions. See Figure 57.

(f) *Genus Sporotrichum* Compact white colonies; requires high $a_w$. Spoilage: stored chilled meats. See Figure 58.

(g) *Genus Cladosporium* Common. Dark colonies. Spoilage: green rot of fruit, vegetables; black spot of meat, eggs, cheese. Wide temperature range for growth, favouring low temperatures.

(h) *Genus Alternaria* Dirty green mycelium, brown multicellular conidia. Spoilage: fruit and vegetables. See Figure 59.

(i) *Genus Fusarium* Produce sickle-shaped multicellular conidia. Spoilage: rot fruit and vegetables; cause discolouration in butter. See Figure 60.

### Yeasts in spoilage

Yeasts tend to grow in acid conditions and where the sugar concentration is high (see page 127). They grow both in aerobic and anaerobic conditions. Fermentative yeasts break down sugars to produce carbon dioxide, alcohols and acids. Oxidative or film yeasts oxidize sugars, organic acids and alcohol, and in their growth raise the pH; they tend to grow on the surface of liquors forming a skin or film. Osmophilic yeasts tolerate conditions of low $a_w$ and are associated with the spoilage of dried fruits, honey, concentrated fruit juices and so on. Salt tolerant yeasts may contribute to the spoilage of brines and salted foods.

*Yeasts important in spoilage*

1  **Saccharomycetales** – true yeasts.

(a) *Genus Saccharomyces* In addition to their many industrial uses some strains which are fermentative and osmophilic are spoilage organisms; for example *Saccharomyces rouxii, Saccharomyces mellis* – spoilage of jams, syrups, pickles, brines, and alcoholic beverages.

(b) *Byssochlamys fulva* This is a 'mould' which also produces ascospores and is classified with the true yeasts. The ascospores are heat resistant causing spoilage of canned fruits.

2  **Cryptococcales** – false yeasts.

   (a)  *Genus Candida*. Some are 'film' yeasts, some of which are acid tolerant, and some osmophilic. Spoilage: high acid foods and brines; some fats such as butter and margarine are attacked by lipolytic strains.

   (b)  *Genus Rhodotorula* Spoilage: 'spotting' of meats.

   (c)  *Genus Torulopsis* Some of these are fermentative and some are salt tolerant. They may cause trouble in brewing.

*Bacteria in spoilage*

Bacterial spoilage starts when vegetative cells are able to grow because the food provides a suitable nutrient environment and the physical qualities of the medium permit growth. Bacteria cause spoilage under many conditions, the main limiting factors to bacterial spoilage being the availability of water and the pH. They require a high $a_w$ and cannot therefore contribute to the spoilage of dry foods while they remain dry.

*Bacteria important in food spoilage*

1  **Gram positive aerobic or facultatively anaerobic cocci:** In regular or irregular groups. Some form non-water soluble pigments.

   (a)  *Genus Micrococcus* Widespread, often isolated from dust and water; some are salt tolerant, some thermoduric, and some psychrophilic. Spoilage: salted foods, pasteurized milk, chilled foods. Optimal growth temperature of 25–30°C for most species.

   (b)  *Genus Staphylococcus* Isolated from the skin. *Staphylococcus aureus* causes a wide range of infections and intoxications including food poisoning. *Staphylococcus epidermidis* is also associated with skin infections. Both groups are salt tolerant – optimal growth temperature 37°C but grow well at lower temperatures and can be associated with food spoilage.

   (c)  *Genus Streptococcus* Widespread. Salt tolerant up to 6.5 per cent w/v solution. Facultatively anaerobic. Need complex vitamin rich foods for growth. Ferment sugars to produce lactic acid. *Streptococcus pyogenes* group – pathogens. *Streptococcus lactis* group – used in the manufacture of dairy produce. *Streptococcus faecalis* group – derived from animal intestine (for example *Streptococcus faecalis, Streptococcus faecium, Streptococcus durans*) and associated with the spoilage of raw meat, fresh and pasteurized dairy produce. Wide temperature range for growth of 10–45°C.

   (d)  *Genus Leuconostoc* Fermentative – produce copious capsules and slimes in favourable environments. Found in slimy sugars, fermenting vegetables, milk and dairy products. Some are salt tolerant, some produce distinct flavours in foods due to diacetyl.

2  **Gram positive cocci – anaerobic.**

   (e)  *Genus Sarcina* Large cocci occurring in packets. Isolated from soil and found on grains.

3  **Gram positive rods – non spore forming.**

   (f)  *Genus Lactobacillus* Varying from long and slender straight rods to coccobacilli. Anaerobic or facultative. Ferment glucose to form lactic acid. Lactose not fermented. Aciduric – optimum pH close to 5. Some thermoduric strains with growth optima above 40°C; many mesophilic strains and some psychrophilic strains. Found in dairy products and effluents, grains and meat products, water, sewage, beer, wine, fruits, fruit juices and pickled vegetables.

4  **Gram positive rods – spore forming.**

   (g)  *Genus Bacillus* Aerobic, occur in dust and soil. Some are mesophilic, such as *Bacillus subtilis*; some are thermophilic for example *Bacillus stearo-*

*thermophilus, Bacillus coagulans* with growth optima 37–55°C. All are active biochemically: saccharolytic, proteolytic and lipolytic strains. Spoilage: aerobic and micro-aerophilic – some strains cause flat sours in canned foods; some saccharolytic strains cause rope, for example *Bacillus subtilis* in bread.

(h) *Genus Clostridium* Anaerobic. Habitat – soil, organic material, animal bowel and excreta. Saccharolytic and putrefactive. The thermophilic species are of importance in spoilage of foods stored at high temperatures. The mesophilic species are important in canning for example *Clostridium botulinum*. Some are proteolytic and putrefactive – for example *Clostridium histolyticum, Clostridium sporogenes*; some are saccharolytic – for example *Clostridium butyricum, Clostridium perfringens*.

5 **Gram negative aerobic rods:** non-sporing. Produce water and non-water soluble pigments.

(a) *Genus Pseudomonas* Widely distributed, biochemically active, aerobic. Source – soil, fresh and sea water, decomposing organic material. Tend not to use carbohydrates but grow well in proteinaceous foods with the production of slime, pigments and odours. They prefer a high $a_w$. Many are psychrotrophic although the optimum temperature for growth ranges between 15 and 40°C. Spoilage: meat, fish, poultry, eggs.

(b) *Genus Halobacterium* Obligate halophiles spoiling foods high in salt.

(c) *Genus Acetobacter* Oxidize ethyl alcohol to acetic acid. Spoilage: alcoholic beverages.

(d) *Genus Alcaligenes* From manure, soil, water, dust. Produce alkaline reaction in some foods. Slimes.

(e) *Genus Flavobacterium* Pigmented colonies – yellow to orange shades, some psychrophilic. Spoilage: discol-

ouration of shellfish, butter, eggs, milk.

6 **Gram negative facultative anaerobic rods.**

(f) *Genus Escherichia* Derived from soil or from the intestine. Their presence in food can indicate faecal pollution. Some species spoil food, fermenting the carbohydrate to acid and gas, and also causing 'off' odours.

(g) *Genus Erwinia* Plant pathogens – causing rots.

(h) *Genus Serratia* Cause red colourations in foods.

(i) *Genus Shigella, genus Salmonella, . genus Proteus* Pathogenic organisms which may be carried by foods.

## Food qualities

Of all the organisms present in a food only certain species will be able to grow and spoil it. The physical qualities of the food, listed below, determine which those organisms are.

1 Water content
2 pH
3 Temperature
4 Gaseous conditions
5 Texture
6 Nutrients

In practice this means that foods can be classified according to their tendency to spoil. If the physical characteristics of the food change, or are changed, it is probable that the potential of the food to spoil will alter (see Table 19).

In order to understand the principles of spoilage, each of the physical qualities of the food must be considered in turn.

### Water content

The more water that a food contains, the more likely it is to spoil and conversely, the less water that a food contains the less likely it is to spoil. Factors which alter the *water status* of a food alter its *spoilage status* too. For example – a moist food is liable to be spoiled by a wide range of organisms. If that food is dried, salted, deep-frozen or preserved in sugar, its water

Table 19   *Food spoilage potential*

| Food category | Examples |
|---|---|
| 1 Non-perishable | Foods in stable preservation – deep-frozen, canned foods (excluding pasteurized ones), sugars, jams, syrups, dry foods. |
| 2 Semi-perishable | Root vegetables, brined and salted foods, semi-moist foods. |
| 3 Perishable foods | Fresh fruit, vegetables, meat, fish, dairy produce – all foods with high moisture content, or held at high relative humidity. |

Table 20   *Examples of foods with high water content*

| Food | gm water/100 gm food |
|---|---|
| Milk, fresh skimmed | 90·9 |
| Milk, fresh whole | 87·6 |
| Eggs, fresh, whole | 74·8 |
| Fruit, grapefruit | 90·7 |
| oranges | 86·1 |
| apricots, raw | 86·6 |
| peaches, raw | 86·2 |
| apples, raw (English eating) | 84·3 |
| pears, raw | 83·2 |
| Vegetables – tomatoes, raw | 93·4 |
| cabbage, raw | 90·3 |
| beetroot, raw | 87·1 |
| peas, fresh, raw | 78·5 |
| potatoes, old, raw | 75·8 |
| Meat – chicken breast, raw | 74·4 |
| beef steak, raw | 66·7 |
| chicken, boiled | 63·4 |
| beef, corned, canned | 58·5 |
| beef steak, stewed | 57·1 |
| Bacon, dressed carcase, raw | 48·8 |
| Fish – herring, raw | 63·9 |
| salmon, smoked | 64·9 |
| salmon, fresh | 68·0 |
| cod, fresh fillets | 82·1 |
| lemon sole | 81·2 |
| Shellfish – crab (boiled) | 72·5 |
| lobster (boiled) | 72·4 |
| shrimps (boiled) | 62·5 |
| oysters (boiled) | 85·7 |
| mussels (boiled) | 84·1 |

*Source*: A. A. Paul and D. A. T. Southgate, *McCance and Widdowson's 'The composition of foods'*, 4th edition of Medical Research Council Special Report, no. 297 (HMSO 1978).

*Note*: The analyses refer to the analysis of the edible portion of the food, i.e. excluding parts such as the skin of fruit, bones in fish and meat.

status alters and it becomes less liable to certain categories of spoilage and more liable to others.

Foods which have a *high water content* are very perishable and likely to spoil rapidly because micro-organisms grow best when water is plentiful. The conditions of storage together with autolytic changes may increase the levels of available water, the quantity and location of which will determine which organisms are able to grow and cause spoilage. Table 20 gives examples of foods with high water contents.

If the moisture content of the atmosphere surrounding a food is in equilibrium with the moisture in the food neither will lose water to the other – a point known as the ERH (equilibrium relative humidity). If they are in a state of imbalance the water content of the food will alter. In a dry atmosphere a moist food will tend to dry out – for example the drying of bread and cakes – whereas in a moist atmosphere a food may absorb water with the result that its water content increases.

Local moisture levels can be altered – for example a sealed polythene packing will not allow water vapour to escape from the atmosphere surrounding the food, with the result that the food surfaces can become very moist and liable to spoilage. The practice of putting hot food into a refrigerator causes condensation which will affect the surfaces of cool foods resulting in risk of surface spoilage of those chilled foods by moulds, yeasts or bacteria.

These vulnerable foods with high water contents can be protected from spoilage in several ways. The reduction of their water content has the effect of decreasing the number

Table 21  *Examples of foods with low water content*

| Food | gm water/100 gm food |
|---|---|
| Cereal foods | |
| biscuits, cream crackers | 3·5–5·0 |
| biscuits, sweet mixed | 3·5–5·0 |
| oatmeal, raw | 8·9 |
| spaghetti, macaroni | 12·4 |
| flour, English | |
| (100% whole wheat) | 15·0 |
| Milk, dried whole | 1·3 |
| Eggs, dried | 5·0–7·0 |
| Fruit and vegetables | |
| dried vegetables | 14·0–20·0 |
| dehydrated fruits and vegetables | 5·0 |
| dried fruits | 18·0–25·0 |
| butter beans, raw | 11·6 |
| beans, haricot | 11·3 |
| Nuts | |
| almonds | 4·7 |
| barcelona | 5·7 |
| brazil nuts | 8·5 |
| peanuts | 4·5 |
| Meat | |
| dehydrated meat | 7·5 |

*Source*: A. A. Paul and D. A. T. Southgate *op. cit.*

of types of organism which can grow in or on the food at its new, lower $a_w$.

The limiting $a_w$ for growth of some groups of spoilage organisms is shown below:

| | |
|---|---|
| Normal bacteria | 0·91 |
| Normal yeasts | 0·88 |
| Normal moulds | 0·80 |
| Halophilic bacteria | 0·75 |
| Osmophilic moulds | 0·62 |
| Osmophilic yeasts | 0·60 |
| No organisms proliferate | 0·50 |

### Dry foods

Dry foods are classed as 'non-perishable' because their water content is too low to allow much microbial growth. Table 21 lists a range of dry foods.

Table 22  *Flora of some dried foods*

| Food | Flora |
|---|---|
| Grain | Natural flora plus organisms from soil and other sources – wide range of bacteria, yeasts, moulds and their spores. |
| Dried fruit and vegetables | Natural flora plus contaminants from processing and retailing. Many spores. |
| Dried egg | Organisms from faecal material on shells; organisms from handling – hundreds to millions per gram. Salmonella can be present. |
| Dried milk | Organisms from udder and from handling; heat resistant strains survive pasteurization – for example species of *Streptococcus*. |

Dry foods are not usually sterile but their microbial flora is unlikely to be able to grow while the dry conditions persist (see Table 22).

There are some organisms known as *xerophytes* which are able to grow at very low moisture contents – these are usually moulds. Prolonged storage under dry conditions usually results in a decline in the number of viable cells and spores present. On the reconstitution of the dried food viable cells are able to multiply and spores are able to germinate. Spray-drying, drum-drying and freeze-drying are methods of preserving foods which tend to allow the deposition of a protective layer of food material around the organisms present enabling them to survive better than unprotected cells.

If a dried food is held under humid conditions it is able to absorb water at its surfaces and will be able to support the growth of moulds; if the water absorption continues to rise yeasts and bacteria will be able to grow. For example when flour becomes damp it is first liable to mould spoilage, the flour amylases convert the starch to sugars and as the $a_w$ rises acid formers such as members of the genus *Acetobacter*, coliforms, lactic acid bacteria and aerobic spore-bearing

Table 23   *Salt preferences of organisms*

| Organisms | Range of tolerance |
| --- | --- |
| Non-halophilic organisms | up to 2% |
| Slightly halophilic | 2–5% |
| Moderately halophilic | 5–20% |
| Extremely halophilic | over 20% |

*Note*: At 10°C saturated salt solution (NaCl) contains 26·3 gm/100 gm.

bacteria, or alcohol formers such as yeasts grow. The flour develops distinctive odours and suffers changes in appearance and quality.

*Deep-frozen foods*
Deep-frozen foods do not support the growth of micro-organisms because the water which they contain is not available while in the crystalline state. The process of freezing tends to reduce the number of viable cells but leaves a significant number which will become reactive when the food is thawed out.

There are many foods which are protected to a certain extent from spoilage by the addition of either sugar or salt in high concentrations. In this way the $a_w$ of the food is lowered. In high concentrations these solutes exert a high osmotic pressure which has an adverse effect on most organisms and destroys them by the withdrawal of water from their cells. A few strains of organism can withstand the osmotic pressures or even prefer these conditions and contribute to the spoilage of foods otherwise protected.

*Salted foods*
The addition of salt in high concentration to foods protects them from spoilage by a wide range of organisms but halophilic organisms – with a preference for such conditions – can cause trouble. Table 23 indicates the salt preference of organisms.

If too little salt has been added to a 'salted' food such as bacon or salt fish, spoilage can result from the growth of both salt sensitive and slightly halophilic organisms. Conversely, in foods where moderate quantities of salt are required in the manufacture, the addition of excess salt can suppress the growth of the desired moderate halophiles and allow the growth of spoilage organisms such as halophilic slime producing yeasts.

The salt itself can introduce halophilic spoilage organisms – the pink slime which is accompanied by unpleasant odours in wet salted fish is due to the growth of pigment producing strains of *Serratia*, *Sarcina*, and *Micrococcus*

Wet fish in the holds of trawlers at sea can be spoiled by halophiles derived from the sea water; and brining solutions for smoked fish can become colonized by halophiles which have an adverse effect on the keeping quality of the smoked product. Examples of salted foods are:

1   *Vegetable preservation* for pickle production 8–11 per cent w/v NaCl. This inhibits spoilage organisms but allows the growth of the desired fermentative lactobacilli and *Aerobacter*.
2   *Ham and bacon* – sides are pumped with pickle – 25 per cent w/v salt which contains a proportion of sodium and potassium nitrate, and sodium and potassium nitrite. It is then dry salted (sprinkled with salt), or tank cured during which time the meat takes up the salt. The final product must not contain more than 500 ppm sodium or potassium nitrate, or 200 ppm sodium or potassium nitrite.
3   *Fish* is either dry salted or brined during which it takes up salt. The final salt concentration in the fish may reach 8 per cent.

*Sweet foods*
In a similar way to salt, sugar is used to protect foods from spoilage. To reduce the $a_w$ to a protective level, very high sugar concentrations are required – nearly 70 per cent, as in jams, preserves, syrups and dried fruits. However osmophilic organisms may be able to grow and

cause spoilage. Moulds such as strains of *Aspergillus* and *Penicillium* tend to grow on the surface of the affected foods. Yeasts such as strains of *Candida, Saccharomyces* and *Rhodotorula* ferment the food with the production of alcohol and $CO_2$ and bacteria can produce ropiness (for example *Leuconostoc* species in syrups), cloudiness (strains of *Bacillus*) or colouration (*Pseudomonas* species) in the affected foods. Examples of foods containing high levels of sugar are: preserves with nearly 68% w/v sugar (jam, curds, mincemeat, fruit fillings), syrups, honey.

*pH*

The pH value of a food limits the range of organisms which it can support (see Table 24 for the pH values of some foods). Foods may be spoiled by the growth of a wide range of organisms whose effect is to alter the flavour, texture and appearance of the affected item, and at the same time pave the way for the growth of other species. Foods manufactured by fermentative processes may be spoiled by contaminating, acid producing organisms during the growth of which unwanted acids and other substances with distinctive flavours are evolved.

A food which has a neutral pH can be protected from some spoilage by neutrophiles by decreasing its pH to such an extent that only a limited range of acidophiles could grow if present. It has long been the practice to preserve perishable foods by the addition of acetic acid (vinegar) – for example, onions, gherkins and pickles. Canned foods can be subject to microbial spoilage – the type of spoilage is dependant, after initial heat processing, on the pH of the contained food.

*Neutral foods*

Examples of the spoilage of neutral foods in which acid is produced and the pH falls are given below. (See Table 24.)

*Meat* (pH 7·0–5·5) may be spoiled by the anaerobic growth of lactobacilli on its surfaces – resulting in the development of surface slime and a drop in surface pH. Under aerobic conditions proteolytic organisms such as pseudo-

Table 24   *pH values of some foods*

| Food | pH value |
| --- | --- |
| lemons | 2·2 |
| vinegar | 2·9 |
| gooseberries | 3·0 |
| prunes, apples, grapefruit | 3·1 |
| rhubarb | 3·2 |
| apricots | 3·3 |
| strawberries | 3·4 |
| peaches | 3·5 |
| raspberries | 3·6 |
| oranges | 3·7 |
| cherries | 3·8 |
| pears | 3·9 |
| tomatoes | 4·2 |
| bananas | 4·6 |
| egg albumin | 4·6 |
| carrots | 5·0 |
| cucumbers | 5·1 |
| cabbage | 5·2 |
| bread | 5·4 |
| meat, ripened | 5·8 |
| tuna fish | 6·0 |
| potatoes | 6·1 |
| peas | 6·2 |
| egg yolk | 6·4 |
| milk | 6·6 |
| shrimp | 6·9 |
| meat, unripened | 7·0 |
| egg white | 7·0–9·0 |

*Note*: The minimum process applied to canned foods with pH values above 4·5 must provide an adequate safeguard against the risk of botulism.

monads may attack the meat protein and release strong smelling compounds associated with spoilage, and as a result the pH of the affected areas may rise.

*Milk* (pH 6·6) contains a range of lactose fermenting organisms which can sour milk rapidly if it is held between 10 and 37°C. Pasteurization destroys the more heat sensitive strains of organisms, leaving the heat resistant lactobacilli and streptococci which are also capable of souring the warm milk. Under chilled conditions pasteurized milk sours slowly because of the reduced ability of these organisms to produce acid at lower temperatures. Later proteolytic organisms may attack the clotted milk proteins and the pH may rise.

### Acid foods

Acid foods (see Table 24) can be spoiled by *aciduric* organisms. An example of this is in the spoilage of wine and beer (wine pH 3·5–4·0) which are manufactured by yeast fermentation processes producing alcohol from carbohydrate. They can be spoiled by the growth of lactobacilli producing lactic acid, members of genus *Aceto-bacter* producing acetic acid, and other fermentative aciduric organisms such as wild yeasts, *Clostridium butyricum* and members of genus *Achromobacter*, producing a whole range of end products.

More commonly acid foods are spoiled by *acidophilic* organisms.

Fruit and vegetables have low pH values – fruits range from pH 2·0 to 4·2 and vegetables, slightly higher – from 4·5 to 7·0. The more acid fruits and vegetables are liable to be spoiled by acidophilic organisms – moulds such as species of *Penicillium*, *Rhizopus* and *Aspergillus* which grow on their surfaces. They may be rotted by saprophytic bacteria such as the slime producing members of genus *Erwinia*, for example, *Erwinia carotovora*, and members of genus *Pseudomonas* such as *Pseudomonas marginalis*. Those acid fruits which contain increasing amounts of sugar as they ripen may be fermented by wild yeasts – damsons may be spoiled in this way.

### Canned foods

Canned foods are divided into two groups based on their acidity. The separation of the two groups is based on the fact that the dangerous food poisoning organism *Clostridium botulinum* is inhibited below pH 4·5 (see page 69). Thus foods which have a pH below this – that is are *more* acid – do not need to be as rigorously processed as those with a pH above 4·5 (*less* acid) in order to be safe from this hazard. Canned foods can be classified into two groups:

*Group 1*   Medium and low acid foods – pH 4·5 and above. For example, meat, sea-foods, milk, meat and vegetable mixtures.

*Group 2*   Acid and high acid foods – pH below 4·5. For example, tomatoes, fruit, pickles, citrus fruits, rhubarb.

### Group 1

Because of the risk of toxin production in the low acid, anaerobic conditions inside the can, medium and low acid foods are processed by heating to above 100°C. Reference to page 120 indicates the exposure necessary to destroy the spores of *Clostridium botulinum*. The type of spoilage to which such foods are subject is due to the survival of spores of the genera *Bacillus* and *Clostridium* which have a greater heat resistance than those of *Clostridium botulinum*.

Thermophilic strains of these two genera grow well at 55°C and only with difficulty at 37°C. Their heat resistant spores can germinate in the canned foods and cause spoilage *if* the can is stored under warm conditions. For this reason they do not often cause spoilage in this country as it is rare for ambient conditions to favour their growth. In hot countries they frequently cause trouble. This type of spoilage can be avoided by the storage of canned foods under cool conditions. *Bacillus stearothermophilus* and *Bacillus coagulans* ferment carbohydrates and cause flat sours; saccharolytic strains of *Clostridia* not only produce acid but also copious quantities of gas (carbon dioxide and hydrogen) with the result that the can swells and distorts. Those *Clostridia* which produce hydrogen sulphide may give rise to 'sulphur stinker' spoilage – which is, fortunately, rare.

Mesophilic bacilli such as *Bacillus subtilis*, *Bacillus cereus*, *Bacillus megaterium* and *Bacillus polymixa* may cause spoilage of canned milk, fish or cured meat causing either flat or gassy sours.

When large hams are canned they are normally only pasteurized – treatment which is inadequate to destroy the spores of *Clostridium botulinum*. The curing salts are inhibitory but only if the can is kept under chilled storage will the product be both safe and unspoiled. Its shelf life is, however, short. Other mesophilic *Clostri-*

*dia* can spoil canned foods – for example the putrefactive *Clostridium sporogenes*; and the saccharolytic *Clostridium butyricum*.

## Group 2

These foods do not require the rigorous processing to which the low and medium acid foods are subjected when canned. Because of this it is possible for heat resistant aciduric bacteria, yeasts and moulds to survive processing and to grow inside the can. For example *Clostridia* and *Bacilli* can cause gaseous spoilage and flat sours in canned fruits. The non-sporing lactobacilli can ferment the foods and strains of *Leuconostoc* can cause slime and ropiness. This type of spoilage can be avoided by slightly more rigorous heat treatment of the product.

The mould *Byssochlamys fulva* being heat resistant can cause spoilage of canned soft fruits, causing them to go soft and pulpy. Exposure to 90°C for a few minutes is an adequate heat treatment.

### Temperature

Any non-sterile food is liable to spoil in time if it is held between the temperatures of −5°C and +70°C – the holding temperature dictating to a certain extent the type of spoilage – although other factors such as the pH and the water content obviously exercise considerable influence. Every type of organism is able to grow within a certain temperature range, maximum growth occurring at and around the optimum temperature. So organisms are divided on the basis of their temperature preferences into rough categories – psychrophiles, mesophiles and thermophiles. Moulds and yeasts tend to grow best at room temperature and below, and therefore assume greater importance in the spoilage of foods held at cool and chilled temperatures. The holding temperature range favours the growth of certain species of the entire flora present – that is those whose optimum for growth is near to the holding temperature. Provided that other environmental factors will allow growth, foods will be subject to spoilage as shown in Table 25.

### Spoilage of meat

Meat originates from warm-blooded animals whose flora is predominantly mesophilic – for example, micrococci, members of the Enterobacteriaceae, bacilli, streptococci, etc., all of which can contribute to the development of surface slime in meats held at room temperature. *Clostridia*, as well as some of these others, can spoil meats held under anaerobic conditions for instance, minced meat, or large volumes of cooked meat, or in vacuum-packed meats – such spoilage resulting in souring and putrefaction. Generally the flora of raw meat is not restricted to those derived from the animal but also includes those derived from the environment, many of which are psychrophilic – for example, members of the genera *Pseudomonas, Achromobacter, Lactobacillus, Bacillus* and others which cause slime and putrefaction of meats stored under chilled conditions. Psychrophilic moulds such as species of *Penicillium, Cladosporium, Thamnidium, Mucor* and others may contribute to spoilage under chilled conditions giving 'stickiness' and 'whiskers'.

### Spoilage of fish

The flesh and body fluids of fish, are generally sterile but the external surfaces and gut harbour large numbers – $10^2$ to $10^6$ per cm$^2$ and $10^3$ to $10^7$ per ml of gut fluid respectively – mostly of species of *Pseudomonas, Achromobacter, Flavobacterium* and spore-bearing anaerobic organisms. Many of these organisms are psychrophilic. The fish are gutted at sea and then either stored on ice or deep-frozen. Fish stored on ice will not keep in first class condition for more than one week because psychrophilic organisms can grow at the flesh temperature of −0·5°C, with the result that after 15 to 16 days their numbers are very high indeed and the fish spoiled. Fish, in good condition, on landing may be contaminated by mesophilic organisms derived from dust, equipment, etc. and be subject to rapid spoilage by them if the fish is sold or awaits processing under conditions of inadequate refrigeration. Deep-frozen fish does not spoil until it is thawed – after which it spoils rapidly.

Table 25    *The storage conditions of foods which affect the type of micro-organisms which can grow and cause spoilage*

| Food | Conditions of storage | | |
|---|---|---|---|
| | Chilled storage | Non-chilled ambient winter, shade | Warm storage very hot weather, heated cabinet, etc. |
| Fresh meat, fish, vegetables, milk, etc. ($a_w$ 1·0–0·95) | Psychrophiles grow | Mesophiles grow | Mesophiles grow, thermophiles may grow |
| Dry foods – noodles, whole egg powder, biscuits, etc. ($a_w$ below 0·5) | | Too dry to spoil | |
| Salted or sugared | Psychrophilic osmophiles and halophiles may grow | Mesophilic osmophiles and halophiles may grow | As 'non-chilled' but spoilage more rapid |
| Acid foods | | Acidophilic organisms may grow | |
| Canned foods | Pasteurized canned hams can spoil slowly | Mesophilic or thermophilic spoilage is possible if improperly processed | |

*Note*: The terms psychrophiles, mesophiles and thermophiles relate to bacteria, yeasts and moulds.

### Spoilage of milk

Mesophilic contaminants in milk derived from the cow are inhibited by chilling, but their slow growth will spoil the milk if it is kept for a few days. Together with these, organisms from the environment will cause it to sour; later when the curd and whey have separated proteolytic moulds and bacteria will break down the protein and the fat will be attacked by lipolytic organisms.

### Spoilage of fruit and vegetables

The flora of fruit and vegetables is derived from the soil and from their own surfaces and comprises bacteria, yeasts and moulds. These organisms tend to possess temperature optima close to those of the normal ambient temperature – in the UK they tend to grow best in the temperature range of 15 to 30°C at which spoilage is most rapid.

### Spoilage of dry foods

Dry foods, are not normally liable to spoilage, but if they become damp they are more prone to do so. Consequently if their storage temperature is 'warm' (above 15°C) any contained organisms will begin to grow. For this reason and because warm dry foods are liable to attack by insects, it is safer to store them in a cool place and under conditions of low relative humidity.

### Spoilage of other types of food

Spoilage of salted and sugared foods is more rapid at higher temperatures, while with thawing, deep-frozen foods become more liable to spoil. Slow thawing of frozen foods may give contained spoilage organisms the opportunity to grow in the thawed outer regions while the inner ones remain frozen.

Acid and canned foods also spoil more rapidly the higher the temperature.

### Gaseous conditions

The oxygen tension and the oxidation-reduction (O–R) potential of a food influences the type of organisms which can grow in it. Spoilage by aerobic organisms occurs at the surfaces of foods; most fresh plant and animal foods have a low O–R throughout and they are aerobic at the surface only. Heating destroys this, as oxygen can diffuse into the food more easily and the food is the more readily spoiled. Facultative organisms grow both on the surface of the foods and within them, as is shown when cans of food are spoiled by members of the genus *Bacillus*. Anaerobic organisms grow within foods held under anaerobic conditions such as inside cans; similarly vacuum-packs are liable to fermentative spoilage by bacteria and yeasts.

### Texture

Fluid foods spoil rapidly because the organisms can easily spread throughout the food by means of their own motility or by convection currents. Semi-solid foods such as meat stews, soup and tinned fruits can spoil as rapidly as fluid foods. Solid foods tend to spoil from their outside surfaces inwards, these being the first surfaces to become contaminated. In other words, the time taken for the whole food to become spoiled depends on how easily the organisms can penetrate the food – thus mincing or otherwise dividing foods aids rapid spoilage because a greater number of surfaces are contaminated by the organisms.

### Nutrients

All foods suitable for human consumption are liable to be spoiled by some micro-organisms. The composition of the food limits which spoilage organisms it can support – most foods supporting a variety. Protein foods such as meat, fish, and eggs are liable to be attacked by proteolytic organisms such as members of the genera *Pseudomonas, Achromobacter, Flavobacterium*; 'carbohydrate' foods such as bread, flour, pasta, syrups and jams are more liable to attack by fermentative organisms; fats are liable to be attacked by lipolytic organisms.

## Summary

The type of spoilage to which a food is subject is determined by the types of organisms which it contains and which of these organisms it is able to sustain. Growth on the other hand is influenced by the physical properties of the food – water content, pH, temperature, gaseous condition, texture, storage conditions and the time over which the food has been stored.

# Chapter 8

# Principles of food preservation

The main aims of the processes used to preserve food are to prolong the time for which the food remains wholesome and to make it safe for consumption. The various methods of preservation enable foods to be kept for periods of time which may vary from 1 to 2 days for pasteurized milk and up to several years for certain canned products. Preservation also allows the import and export of food and its transport from areas of food production to the large centres of population in towns and cities. In addition, certain foods no longer have to be seasonal since after harvesting they may be preserved and distributed at any time throughout the year.

Some preservative methods such as sun-drying have been practised for many centuries, while others, such as accelerated freeze-drying, are of recent origin. Certain processes which are used to preserve foods may also be employed to alter their nature and impart a desired flavour, as in smoking and acid fermentation.

In Chapter 7 it was established that one of the main causes of food spoilage is the growth and activity of micro-organisms; and the methods of preservation which are employed are based on a knowledge of the physical and chemical factors influencing and preventing the growth of these organisms. The more important preservative methods will be discussed and these are classified under the following headings:

1  High temperature preservative treatments
2  Low temperature preservation
3  Dehydration
4  Chemical preservatives
5  Irradiation
6  Controlled atmosphere packaging

Often these preservative methods are also successful in preventing food spoilage due to enzyme catalysed reactions.

## High temperature preservative treatments

When micro-organisms are subjected to high temperatures, the proteins of the microbial cell are denatured, and the enzymes inactivated. Higher temperatures or longer exposures do progressively more damage to each cell and very severe heat may kill the cell through oxidation (burning). The amount of heat required to process a food is dependent on many factors which relate, on the one hand, to the organisms present, and on the other, to the type of food being processed.

The organism related factors are:

1  The species of organism present
2  The numbers of organisms present
3  The heat resistance of vegetative cells
4  The heat resistance of spores present
5  The qualities of the food which affect the microbial heat resistance – such as the pH and the presence of sugars or salts
6  The storage temperature of the processed food

The food itself determines the type of treatment too. It is very important that the correct length of time at the correct temperature is applied in all parts of the food treated. This is more easily achieved in a food of uniform texture – such as milk or fruit juice – than in a food of varied viscosity or density. Heat penetration into the centre of a large can of meat (weighing perhaps 3 kg) is achieved with more difficulty than with a can weighing 500 gm. Care must be taken so that the outside layers of food are not over-cooked. Overheating changes the flavour of a heat-treated food and should be avoided while at the same time ensuring that the contained population of micro-organisms is satisfactorily killed.

The methods of heat treatment which are employed in the processing of food are:

Pasteurization
Sterilization
Cooking (domestic processes)

### Pasteurization

The process takes its name from the Frenchman Louis Pasteur who was the first person to realize that food spoilage was caused by the growth and activity of micro-organisms. It is a term now widely applied to mild heat treatments which reduce the number of spoilage organisms present and kill the pathogenic organisms but does not render the food sterile. Although commonly associated with milk, pasteurization is also used to treat a variety of foods.

Pasteurization is best applied when the target organisms are not very heat resistant; when higher temperatures would harm the food; when additional preservative methods such as refrigeration are used after heat treatment; and when the main object is to kill pathogenic vegetative cells.

The times and temperatures involved in pasteurizing vary from food to food.

### Deciding on a process

If a food contains only one species of heat sensitive organism which is of importance (organism A), the process used would be based on that organism. Its destruction in the food at a range of temperatures and times would be studied and a series of curves would be constructed similar to those shown in Figure 39 (page 42). From those data a second curve like Figure 40 (page 43) would be prepared. This would show how the population of organisms would be reduced to 10 per cent of the original number at a whole range of temperature and time combinations. Then it would have to be decided whether reduction to 10 per cent of the original number was an adequate treatment. This could only be determined if values for the likely numbers of organisms in the untreated product were available.

Figure 61(a) shows a 1D curve which is similar to that shown in Figure 40 (page 43), together with the curves if the population is reduced by 99 per cent (2D), 99·9 per cent (3D), 99·99 per cent (4D) and 99·999 per cent (5D). Which of these describes appropriate treatment?

Suppose the food contains between $10^4$ and $10^6$ organisms per ml when it is untreated; and suppose that no more than $10^1$ organisms per ml would be acceptable in the processed food, then the minimum heat treatment must be a 5D process ($10^6$ ... $10^1$ per ml). Any combination of time and temperature along the 5D line would be the minimum necessary but would be adequate.

If a second organism (organism B) of different heat sensitivity were present in the raw product, would the heat treatment just described be adequate to destroy that organism? Again, new data would have to be accumulated. The same questions would need to be asked – how many of organism B are there in the raw product? How many are acceptable in the processed product? What level of treatment is adequate to destroy these numbers? Figure 61(b) shows possible heat treatments to deal with organism B.

If a third organism (organism C) were present, Figure 61(c) shows the possible heat treatments.

But what if all three organisms were present together in the raw product – but at different population densities? Table 26 gives some possible data, and the acceptable level of organisms after processing.

Which heating process is adequate to satisfy

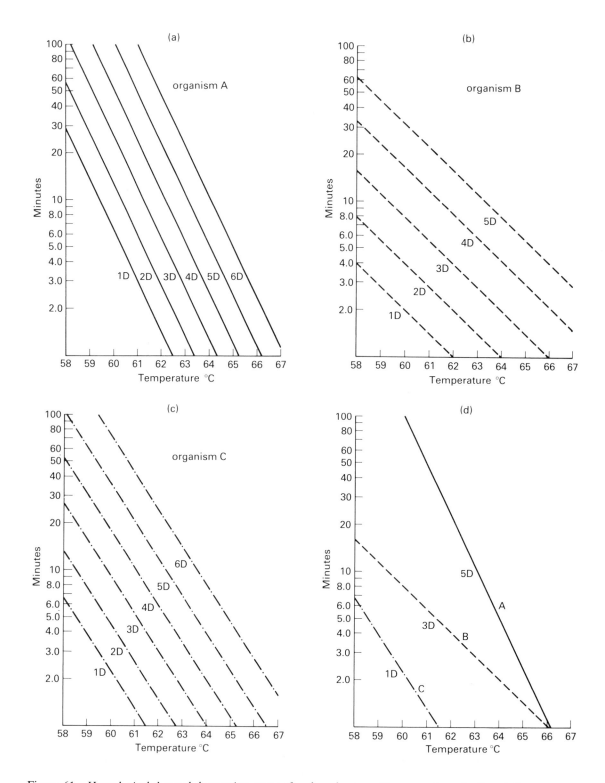

Figure 61 *Hypothetical thermal destruction curves for three heat sensitive organisms A, B and C*

Table 26 *Hypothetical data for organisms A, B and C – their occurrence in untreated product and acceptable numbers in the produced product*

| Organism | Cell concentration in raw product | Characteristics | Acceptable level in processed product | Treatment required |
|----------|-----------------------------------|-----------------|---------------------------------------|--------------------|
| A | up to $1 \times 10^6$ per ml | Spoilage organism | Less than $1 \times 10^1$ per ml | 5D |
| B | up to $1 \times 10^1$ per ml | Pathogen | Less than 1 per 1000 ml | 3D |
| C | up to $1 \times 10^2$ per ml | Spoilage organism | Less than $1 \times 10^1$ per ml | 1D |

the three requirements? Figure 61(d) combines the 5D curve for organism A, the 3D curve for organism B and the 1D curve for organism C. It can be seen that any treatment described by the curve for organism A will adequately meet the criteria set. But some treatments may be impractical – for example sustaining an even temperature of 60°C throughout a food, for 100 minutes is not an easy process. Keeping the food at around 63°C for 11 minutes would be easier. It can also be seen from Figure 61(d) that if a temperature over 66.2°C has to be used for less than 1 minute, to meet the criteria, only the times and temperatures determined by the extrapolation of curve B (and no longer curve A) would be adequate.

However, the treated product would only be satisfactory if the criteria set were based on good original information about the occurrence of organisms A, B and C in the raw product. If it were difficult to obtain such data then safety margins would have to be built in – and a more rigorous treatment applied. If the new criteria were then 6D for organism A, 4D for organism B and 2D for organism C, any of the processes indicated by the grey zone in Figure 62 would be satisfactory, provided that no other deterioration of the product occurred. Possible deterioration as a result of heat treatment might be the destruction of vitamins, browning, or the destruction of a physical property such as the ability in milk to form a cream line.

Additionally, practical considerations are of great importance. While in theory a process of 6 seconds at 74.3°C is adequate (see Figure 62), would it in practice be possible to heat the product for such a short time? How could it be done without running the obvious risk of exposure for a much longer time period and would a longer exposure harm the product?

So there are two more factors which must be considered when designing a process. The first is how is the heating to be achieved? For example, will the product remain stationary and be heated in batches, or can it be pumped through a heating device? The second factor is that all the heat received by organisms contained in a food product, above the maximum growth temperature, has a lethal effect. Process calculations take the cumulative effect of heat into account and the desired level of destruction of organisms may be achieved by final temperatures within the food which are lower than those demanded by theory, since no food can be instantaneously raised to a predetermined temperature.

In the pasteurization of fluids such as milk and fruit juices, the products are free-flowing. They are raised to the pasteurization temperature by a process of pre-heating (usually involving flow through heat exchangers) and then held at the

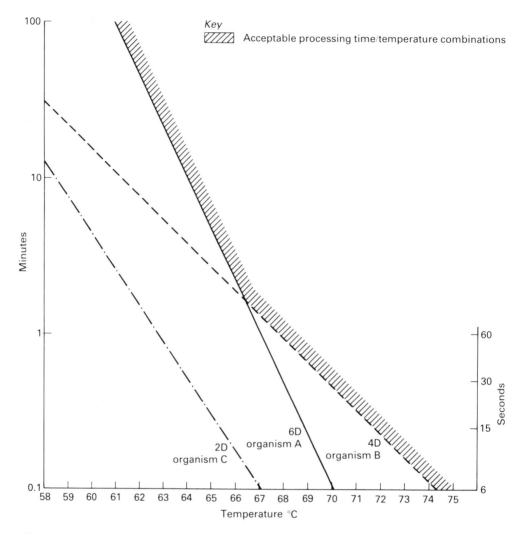

Figure 62  *Acceptable processing time/temperature combinations for raw products containing organisms A, B, C*

*Note*: These data are simply abstracted from Figure 61(a), (b) and (c), and should be contrasted with Figure 61(d).

predetermined process temperature. This is achieved by installing a holding tube of carefully calculated length in the equipment. The flow rate of the fluid is carefully controlled so that it takes a fixed number of seconds for the product to flow along the length of tube. The exit temperature of the liquid will not be less than that required for safe processing. Such a technique permits high temperature, short time processing.

Batch methods of heating involve greater problems of heat transfer, with the risk of outer layers being overheated and central regions being under-heated, although stirring helps to distribute the heat evenly.

*Milk*

The two usual methods for treating milk are: high temperature short time (HTST) which involves a temperature of 71·7°C for 15 seconds and low temperature holding (LTH) the temperature being 62·8°C for 30 minutes.

After being subjected to either of the processes, the milk must be cooled immediately to not more than 10°C. These times and temperatures, laid down in The Milk (Special Designation) Regulations, 1963, are sufficient to reduce the number of spoilage organisms (thereby prolonging the life of the milk) and also to kill any pathogens present. Many organisms which cause infectious diseases have in the past been transmitted by milk and these include the organisms responsible for tuberculosis, diphtheria, scarlet fever and poliomyelitis.

Because heating criteria are defined and required by law, some system of monitoring whether satisfactory pasteurization has been achieved is also needed.

Research shows that an enzyme alkaline phosphatase, present in milk, is progressively destroyed by heating and that the heat required for its destruction is slightly in excess of that required to destroy tuberculosis organisms in milk. The presence of the alkaline phosphatase in milk can be detected by a colorimetric test. Thus the criteria for the heat treatment of milk have been set as those which ensure the destruction of alkaline phosphatase and therefore the tuberculosis organisms and other more heat sensitive organisms. If, on testing, active alkaline phosphatase is detected, it is deemed that the milk has not been satisfactorily heat treated.

*Egg*
Bulk liquid egg has been compulsorily pasteurized in England and Wales since 1 January 1964. Legislation to pasteurize this produce was introduced in order to remove the danger of salmonella food poisoning. Eggs may be contaminated with salmonellae and, when pooled together in the form of bulk liquid egg, there is a danger that one or more contaminated eggs will infect the whole batch. Under The Liquid Egg (Pasteurization) Regulations, 1963, the bulk liquid egg is treated at a temperature of 64·4°C for 2·5 minutes followed by immediate cooling to below 3·3°C. Since these Regulations have

been in operation this product has generally been regarded as safe. Spray dried egg may be heat treated before drying but this does not ensure a salmonella-free product.

If liquid egg is overheated it coagulates. Therefore, any process to pasteurize liquid egg must be mild enough to avoid protein coagulation. The process which is used to pasteurize egg has to take this factor into account as well as the satisfactory level of destruction of the salmonellae. The process also destroys an enzyme normally present (α-amylase) which can hydrolyse starch. So the test for satisfactory pasteurization looks for the absence of active α-amylase – iodine and starch are added to the treated egg and satisfactory treatment is indicated by a black colour due to the combination of starch and iodine. The active enzyme would hydrolyse the starch and the resulting colour would be brown (not black) in the presence of iodine.

*Ice-cream*
Liquid ice-cream mix is another product that is compulsorily pasteurized, as in the past it has been responsible for outbreaks of food borne infections such as typhoid and paratyphoid fevers and also staphylococcal food poisoning. The time and temperatures used vary and are laid down in The Ice-Cream (Heat Treatment etc.) Regulations, 1959. They are:

65·6°C for 30 minutes
71·1°C for 10 minutes
79·4°C for 15 seconds

After treatment the ice-cream must be cooled to 7·2°C within 1·5 hours and kept at this temperature until frozen (see Appendix 1).

Although specific heat treatments are required by law, there is no suitable chemical method of telling whether satisfactory pasteurization has been achieved. However, a test based on bacterial activity – the methylene blue test – which is not precise enough for statutory purposes, is the recommended test for use by local authorities to measure the effectiveness of the treatment.

Other products which may be pasteurized, mainly to reduce the number of spoilage organisms, include beer, citrus fruit juices, vinegar and dried fruits, although there are no statutory standards for these products.

### Sterilization

In order to sterilize most foods, temperatures above 100°C are required. These temperatures are obtained by means of steam under pressure, as for example in a pressure cooker or an autoclave. When deciding the times and temperatures used in sterilizing procedures such as canning, one must not only take into account the heat resistance of the micro-organisms but also the effect of the process on the food. It is possible to obtain a sterile product by prolonged heating but in the meantime it may have developed odd flavours and become mushy or discoloured. Other important factors that have to be considered include the size of the can, the acidity of the food and the heat penetration achieved through the food.

### Milk

Sterilized milk is not necessarily free from all micro-organisms but all vegetative cells and the majority of spores are killed. In one common method of sterilization, the milk is homogenized at 66°C, placed in capped bottles and heated at temperatures ranging from 105°C to 110°C for between 20 and 40 minutes. This procedure prolongs the keeping time for at least 7 days and usually longer but unlike pasteurized milk there is a change of flavour and appearance.

The so called long-life milk is produced by the ultra-heat treatment process (UHT) during which the milk is held at 135°C for about 2 seconds. This process kills all vegetative cells and spores present. After the heat treatment the milk is aseptically placed in sterile containers and provided that it is not recontaminated during this procedure, the milk can be kept unopened for 6 months without any deterioration in quality. This heat treatment has the advantage of producing very little change in flavour or loss of nutrient quality.

### Canning

The use of canning as a method of preserving

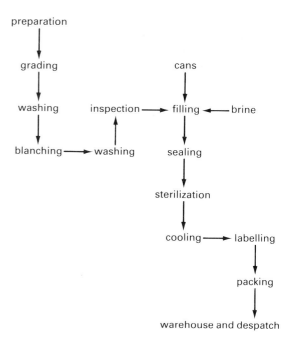

Figure 63  *A flow-line diagram of one method of manufacture for canned processed peas*

foods goes back to the end of the eighteenth century when Nicholas Appert discovered that when food was placed in hermetically sealed jars and heated, it remained wholesome. Since his observations, canning has become universally accepted as a safe method of preservation and is the most important in terms of tonnage and variety of foods.

Food before canning undergoes a variety of treatments depending on the food to be canned. Vegetables, for instance, are graded, trimmed, washed and blanched; the latter process inactivates the enzymes present, removes gas from the tissues causing some shrinkage and also softens the tissues enabling more to be placed in the cans without damage. (See Figure 63.)

The cans are filled, heated, hermetically sealed and then processed. The time and temperatures used in canning procedures vary but they must be sufficient to ensure that all pathogens are either inactivated or destroyed. The most heat resistant pathogen one may expect to find in canned food is *Clostridium botulinum*, although there may be non-pathogenic spore-forming spoilage organisms present which are even more heat resistant, an example being *Bacillus stearothermophilus*. Destruction of the spores of this heat resistant spoilage organism will also ensure the destruction of *Clostridium botulinum* spores.

The term 'sterile' denotes complete destruction of micro-organisms. A canner, during processing, aims at producing a sterile product but does not always achieve this, as in some instances it may be impractical. However the small number of heat resistant spores which are left in these cases are not able to grow under the conditions present within the can. For example, the storage temperature may be too low if they are thermophilic spoilage organisms such as *Bacillus stearothermophilus* or the contents may be too acidic. The food is then said to be 'commercially sterile'. In the canning industry the term 'commercially sterile' means that the food has been processed sufficiently so that under normal conditions of storage it will not spoil, or harm the consumer.

In determining a heating process for canned food two factors are combined:

1   How many spoilage organisms of heat resistance equal to (or less than) the indicator organism are likely to be present per can (say $1 \times 10^3$ per can)?
2   What level of cans capable of being spoiled can be accepted?

To prevent any one can from spoiling, enough heat must be applied to destroy all 1000 spores present. A 3D process would reduce the 1000 spores to one living spore per can and, potentially, the can could be spoiled as a result of the growth of that single organism. A 4D process would mean that 0.1 of an organism was contained in each can (in effect, one organism in every ten cans). If the spoilage of one can in ten is acceptable, a 4D process would be adequate. However, this is not usually the case – a lower level of failure would normally be demanded. Perhaps a potential level of failure which would be acceptable would be one can in 100,000. This

would mean that with a raw product containing about $10^3$ spores per can, the minimum process which could be applied would be an 8D process. Thus if the D value for the indicator organisms, contained in the particular food product under consideration, was 1 minute at 121°C, then a process involving 8 minutes at 121°C would satisfactorily treat the product. As with pasteurization, the process has to be very carefully calculated and controlled because failure to heat treat all the contents of the can to the required temperature and time would mean processing failure; equally there is the risk of overheating the outside layers of food within a can in order to achieve thorough heating. In addition, the heat applied at a level below that set still has some sterilizing effect, both while heating up and while cooling down and this heat is taken into account during process calculations.

Figure 64 indicates the range of temperatures and times necessary to destroy different, important, indicator spores. The destruction of *Clostridium botulinum* spores is important to ensure that botulism does not occur as a result of their growth in the anaerobic conditions in low acid foods. The processing reduces the contained population of spores of equal or lesser heat sensitivity than *Clostridium botulinum* by a minimum factor of $10^{12}$ (that is, 12D – the 'botulinum cook' = 2·5 minutes at 121°C). This process is insufficient to destroy putrefactive anaerobes of greater heat resistance such as *PA3679* which is often used as an indicator organism for processing to avoid considerable losses through spoilage. Processes based on multiples (4D or 5D usually) of the 1D TDT process are applied as appropriate to foods liable to be spoilt by putrefactive anaerobes.

When thermophilic spoilage is a risk (particularly where storage conditions are liable to be hot), processing may be based on *Bacillus stearothermophilus*, but since its destruction may reduce food quality, the use of very high quality ingredients with low initial spore counts is a more satisfactory form of protection.

Acid and high acid canned foods (see Table 27) undergo a less severe treatment than medium or low acid canned foods, as the

Table 27  *Classification of canned food based on acidity*

| Acidity | Canned food |
| --- | --- |
| Low acid pH 5·3 and above | Meat products, sea foods, marine products, milk and certain vegetables |
| Medium acid pH 5·3–4·5 | Meat and vegetables mixtures, spaghetti, soups, sauces and some vegetables |
| Acid pH 4·5–3·7 | Tomatoes, pears, pineapples and other fruits |
| High acid pH 3·7 and below | Grapefruits, citrus juices, rhubarb, pickles and other fruits |

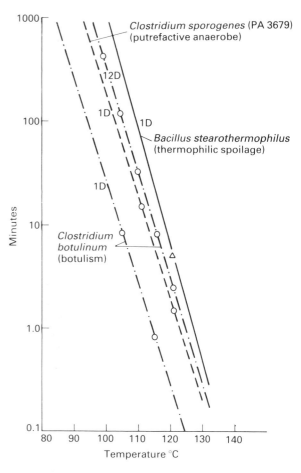

Figure 64  *Thermal death time (TDT) curves for indicator spores important in canning*

micro-organisms capable of causing spoilage at pHs below 4.5 mostly have a comparatively low heat resistance. The possibility of growth and toxin production from *Clostridium botulinum* does not exist as these low pHs are inhibitory to it. In addition, *Clostridium botulinum* spores, like most other bacterial spores, are much more easily destroyed at acid pHs than at neutral ones. The more acid the food the lower the temperature used in processing – for example, canned tomatoes of a pH less than 4·5 require only a few minutes at 100°C. Medium and low acid foods however must be heated sufficiently to destroy the heat resistant spores of *Clostridium botulinum* and conditions drastic enough to kill these spores will also ensure the destruction of all other pathogens. Meat and vegetables are usually processed at 115°C while other foods may be processed at higher temperatures but for shorter times, as in HTST canning. In the latter process the food is sterilized or commercially sterilized outside the can and is aseptically placed in sterile containers which are then sealed. One advantage of this method is that it is possible to shorten the sterilizing time and for many commodities this results in an improvement in quality.

Some types of low acid canned foods such as cured meats, however, are given milder heat treatments than other low acid canned products, as they contain curing salts which also act as microbial inhibitors aiding preservation. Large canned hams and luncheon meats (2·5 kg) are only subjected to temperatures akin to pasteurization and should not be regarded as sterile. It is recommended, however, that after the heat treatment they undergo, there are no viable vegetative cells and the number of heat resistant spores be limited to less than 1000 per gram. The reasons for the fairly mild heat treatment is to keep shrinkage and loss of flavour to a minimum. As they have only been subjected to pasteurization temperatures the hams should be stored in a refrigerator at temperatures below 5°C.

After processing, the cans are cooled quickly under water which causes the water vapour in the headspace to condense, producing a partial vacuum inside the can, so that if there is a faulty seam water may be drawn into the can. Organisms have been known to enter through damaged seams during cooling – for example salmonellae from contaminated cooling water and staphylococci from the hands of the operatives. The cooling water should be treated in some way to ensure that it is bacteriologically safe and there should be careful handling of cans during processing to avoid damage to the seams. Canned foods are safe provided they have been properly processed and not recontaminated as a result of an imperfect seal.

The storage life of canned foods varies; some meat products, for example, may be stored for up to eight years, although their flavour may begin to deteriorate after three years or so due to chemical changes; while certain canned fruits may have a more limited life as the contents may corrode the can.

### Cooking

Although not generally regarded as a method of preservation, cooking of food nevertheless should result in a reduction in its microbial load and may prolong its keeping time. Most of the cooking methods involve heating at about 100°C or below. The outside of the food may reach 100°C or above, as in frying, but the centre does not normally reach boiling point. In the roasting of meat the internal temperature may only reach 85°C, as, for example, in well done pork. The number of organisms destroyed depends on a number of factors which include the length of the cooking process, the temperature used, the nature of the food (heat penetration being faster through a liquid than a solid food) and also the size of the food being cooked. Enzymes may also be inactivated during cooking thus minimizing any chemical changes. (See also page 176.)

## Low temperature preservation

Low temperatures slow down microbial meta-bolism and prevent growth. The lower the temperature the less the activity and, as a result, death of some organisms ensures.

Low temperature preservation may be studied under the following headings:

Cellar storage
Chilling storage
Freezing
Gas storage

### Cellar storage

This does not involve any artificial means of reducing the temperature. The temperature of cellar storage is usually only a little lower than that of the outside atmosphere. The types of product stored under these conditions include root vegetables, cabbages and fruits such as apples. This kind of storage is suitable for stable and some semi-perishable foods. The relative humidity must be carefully controlled as too little moisture will lead to shrivelling of the stored products and too much moisture to 'sweating' which encourages the growth of spoilage organisms.

### Chilling storage

The temperature of chilling storage can vary considerably and the term can be applied to any reduction in the normal temperature of food. The temperature used varies with the nature of the product. Too low a temperature for example can bring about undesirable changes in some fruits, such as the internal browning of apples. The temperature of an ordinary refrigerator is usually maintained below 5°C, that is between 1°C and 4°C, which inhibits the growth of pathogens and many spoilage organisms. Growth of salmonellae, staphylococci and *Clostridium perfringens* will not occur below 5°C but the organisms remain viable for prolonged periods in refrigerated foods and growth may be resumed once the temperature is raised. *Clostridium botulinum* type *E* however has been found to grow and produce toxin at temperatures between 3°C and 4°C, so to ensure safety from this organism, food should be stored below 3°C.

## Freezing

Freezing stops the multiplication of micro-organisms and causes an initial rapid decrease in the number of viable cells which continues at a slower rate during storage in the frozen condition.

Most foods are made up of a large percentage of water which enables them to be frozen and all the water in most foods is in the frozen state at temperatures below −10°C. Before freezing, foods undergo a variety of processes which include sorting, preparing, washing and blanching. Blanching is achieved by dipping the fruit or vegetable, for example, in hot water or by the use of steam. This process, in addition to inactivating the enzymes present and preserving the green colour of vegetables, also helps to reduce the microbial load.

Foods may be frozen slowly in air at −15°C to −29°C, the process taking from 3 hours to 3 days. Large ice crystals are formed which damage the cell walls so that on thawing there is a loss of nutritive substances from the food. Quick freezing is a more satisfactory process as small ice crystals are formed which do not distort and damage the cells. Quick freezing may be defined as any process where the temperature of the food passes through the zone 0°C to −4°C within 30 minutes. The temperatures used vary with the processes utilized and range from −17·8°C to −45·6°C. Two main methods are employed at present, the first involves the product being held between metal surfaces which are cooled by a refrigerant, while the second involves the commodity being subjected to a blast of very cold air in a specially built, insulated tunnel. The first method is more suitable for freezing commodities such as meat and fish which take a fairly long time, while the second method is better for foods such as peas, sliced beans and other vegetables. (See Figure 65.) A third and more recent method of quick freezing is by direct immersion of the product in liquid nitrogen. This method has a number of advantages, one of which is the rapidity of the freezing since the food is in direct contact with the refrigerant. Liquid nitrogen boils at −196°C and boiling occurs when food is immersed in it due to the rapid transfer of heat from the food. Freezing by this method may be complete in 30 seconds. It is, however, expensive but has proved useful for preserving foods which cannot satisfactorily be treated by other means. Strawberries are fruits that are successfully frozen by this method. Foods may also be quick-frozen by dipping them in very cold fluids such as invert sugar solution or brine.

Although a percentage of vegetative cells may be destroyed by freezing, this method cannot be relied upon to eliminate food poisoning organisms and once the food is thawed, growth will be resumed. Freezing has little effect on bacterial spores, while viruses and toxins are also stable in the frozen condition. Between −5°C and −10°C

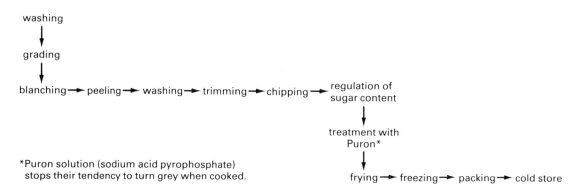

Figure 65   *A flow-line diagram of one method of quick-frozen chip manufacture*

very little microbial growth of any kind is observed. However there are exceptions. Bacterial growth, for example, has been recorded at −20°C while a certain pink yeast has been reported to grow at −34°C. In experiments carried out on food poisoning organisms it was found that salmonellae disappear after one month and staphylococci after five months from 'quick-frozen' strawberries at −18°C. Salmonellae however have been isolated after two years or more of frozen storage in whole egg products and meats.

Frozen convenience foods may either be stored raw or they may be cooked before freezing. If they are cooked first, the heat treatment they undergo must be sufficient to destroy pathogens and most spoilage organisms, as before consumption they may only be warmed. If the foods are frozen raw, hygienic food handling is the main way of reducing microbial contamination and the food should be further protected by adequate packaging. The temperature of freezing storage varies but is usually −18°C to −23°C.

Thawed foods should not be left at room temperature for a prolonged period and neither should they be refrozen as they could be carrying a higher microbial load.

### Gas storage

Gas storage is sometimes used in conjunction with cold storage to retard food spoilage. Carbon dioxide is added to the storage atmosphere in concentrations which vary with the kinds of food stored. For meat such as beef, concentrations of 10 per cent are used, while fruits, such as apples, require between 5 and 10 per cent. The higher the concentration of carbon dioxide the less the spoilage but at concentrations above 20 per cent undesirable changes take place – meat may turn brown for example.

## Dehydration

The water content of the food is reduced to levels below which micro-organisms cannot grow and multiply. The number of micro-organisms in the product will depend on the initial load before dehydration. When water is added to the food the organisms will recommence growth provided the food is a suitable substrate.

In common with canning and freezing, food before drying also undergoes a variety of treatments which in the case of vegetables include washing, grading and blanching. Other foods such as meat may be cooked before being dehydrated.

The oldest method of preserving food is sun-drying but it is limited to countries with hot climates. Foods preserved in this manner include prunes, raisins and apricots. Sun-dried foods, however, are subject to contamination from air borne animal infections as well as from insects and birds.

Because of the limitations of sun-drying the majority of foods are dried by artificial means. Some are dried by placing them on perforated trays in hot air tunnels – for example fruits, vegetables and some meats. A recently developed method (called fluidized bed-drying) involves the passing of warm air through foods in a perforated bed with fans underneath. The particles of food are kept in motion during the process to prevent portions from sticking together. This method has proved particularly useful for drying vegetables. (See Figure 66.) Another method which is widely used is spray drying, which involves the spraying of the products into a rapidly moving current of hot air. The particles of low moisture content drop to the bottom of the drying chamber and are removed. The original product should possess low numbers of contaminating organisms before dehydration because the treatment it undergoes cannot be relied upon to destroy all pathogens. Because these foods are hygroscopic they are placed in protective packaging which helps in preventing recontamination and water absorption. Products which are spray dried include milk and eggs. Moisture may also be removed by roller drying which involves the application of pastes and mashes to the surface of heated drums which are then scraped off as a thin layer. Products dried in this manner include milk, potato flakes, tomato paste, instant breakfast

Tenderometer ⟶ washing ⟶ size grading ⟶ pricking ⟶ blanching
test

fluidized ⟶ hot air ⟶ bin drying ⟶ packing
bed drying    bed drying

The drying process takes place in three stages:

1  Fluidized bed drying. This consists of perforated beds with fans underneath. As the peas drop
   from one bed to another the temperature is increased from 40°C, which is the temperature of
   the initial bed, to a maximum of 55°C, and between the first and the seventh fluidized bed the
   moisture content has been reduced from 80 to 50 per cent.
2  Hot-air bed drying. The moisture level falls to 20 per cent after treatment on these beds, which
   have air passing through from below but no fluidization.
3  Bin drying. The peas are transferred to bin dryers and kept in them until the moisture content
   falls to between 5 and 7 per cent, the whole process taking approximately 16 hours.

Figure 66   *A flow-line diagram of a process used in the production of hot-air dried peas*

cereals and animal foods. Drum dried foods are generally not as satisfactory as spray dried foods as they have a more 'cooked' character and do not reconstitute as well.

### Accelerated freeze-drying (AFD)

Under certain conditions of low vapour pressure, water can evaporate from ice without first going through the liquid phase, that is, it sublimes. In this method the food is first frozen and then placed in a vacuum chamber. Heat is applied to the food to speed up the sublimation. This method has the advantage that the food does not have a chance to distort or shrink and it appears as a spongy mass in the dried form which may be quickly reconstituted. This method is used to dry a variety of foods such as meat, poultry, fish, shellfish, vegetables and fruit. In common with all other drying methods, the degree of contamination after drying depends on the initial load. Few bacteria are killed in this process and therefore the foods must be processed under strict hygienic conditions. There is considerable handling of the product by the operatives so this adds to the precautions that must be taken. Freeze-dried products are stable at room temperature but for long storage they should be refrigerated. As with spray-dried and hot air-dried products, protective packaging is necessary. The products are usually vacuum-packed or packed with an inert gas to prevent oxidation.

Bacteria need more moisture for growth than either moulds or yeasts. Most bacteria including food poisoning organisms require $a_w$ values above 0·90 for growth and therefore are not often responsible for spoilage of dried foods. Staphylococci however have been observed to grow in media with a water activity of 0·86 although their rate of growth is considerably reduced at $a_w$ values below 0·94. It must be remembered that food poisoning organisms and spoilage organisms can survive dry conditions and recommence growth once moisture becomes available. Foods with an $a_w$ below 0·65 seem to be very stable as very few organisms are known to grow at this low water activity.

## Chemical preservatives

Chemicals prevent microbial growth in the following ways:

1  Interference with cell permeability
2  Interference with enzyme activity
3  Interference with the genetic mechanism

### Preservatives governed by the Preservatives in Food Regulations, 1979

The use of chemicals as a means of preserving foods is governed by the Preservatives in Food Regulations, 1979. These Regulations list the permitted preservatives, the foods to which they may be added and specify the maximum

Table 28 *Examples of articles of food which may contain preservatives and the nature and proportion of preservative in each case. Preservatives in Food Regulations, 1979*

| Specified food | Permitted preservative | mg/kg not exceeding |
|---|---|---|
| Beer | Sulphur dioxide | 70 |
| Bread | Propionic acid | 3000 |
| Cheese | Sorbic acid | 1000 |
| Cheese, other than Cheddar, Cheshire, Granapadano, provolone-type cheeses or soft cheese | Sodium nitrate and sodium nitrite | 50, of which not more than 5 may be sodium nitrite expressed in both cases as sodium nitrite |
| Flour, confectionary | Propionic acid or sorbic acid | 1000 1000 |
| Fruit, fresh: | | |
| bananas | 2(thiazol-4-yl) benzimidazole | 3 |
| citrus fruit | Biphenyl | 70 |
| grapes | Sulphur dioxide | 15 |
| Fruit, crystallized, glacé or drained | Sulphur dioxide | 100 |
| Fruit, dried, other than prunes or figs | Sulphur dioxide | 2000 |
| Gelatin | Sulphur dioxide | 1000 |
| Potatoes, raw, peeled | Sulphur dioxide | 50 |
| Sausages or sausage meat | Sulphur dioxide | 450 |
| Snack meal, concentrated with a moisture content of not less than 15 per cent and not more than 60 per cent | Sorbic acid or methyl 1, 4-hydroxybenzoate | 1500 175 |
| Vegetables, dehydrated(other than cabbage, potato or Brussels sprouts) | Sulphur dioxide | 2000 |
| Vinegar, cider or wine | Sulphur dioxide | 100 |

| Specified food | Permitted preservative | mg/kg not exceeding |
|---|---|---|
| Pickles, other than pickled olives | Sulphur dioxide, benzoic acid, Methyl or ethyl or propyl 4-hydroxybenzoate | 100 250 250 |

*Note*: This table does not include all foods or all preservatives listed in Schedule 1 of the Preservatives in Food Regulations, 1979.

permitted levels (see Table 28). In the context of the regulations the word 'preservative' means any substance which is capable of inhibiting, retarding or arresting the process of fermentation, acidification or other deterioration of food or of masking any of the evidence of putrefaction. The Regulations however do not include the following substances:

1 Any permitted antioxidant
2 Any permitted colouring matter
3 Any permitted emulsifier or permitted stabilizer
4 Common salt (sodium chloride)
5 Lecithin, sugars or tocopherols
6 Nicotinic acid or its amide
7 Vinegar or acetic acid, lactic acid, ascorbic acid, citric acid, malic acid, phosphoric acid, polyphosphoric acid or tartaric acid or the calcium, potassium or sodium salts of any of the acids specified in this list
8 Glycerol, alcohol or potable spirits, isopropyl alcohol, propylene glycol, monoacetin, diacetin or triacetin
9 Herbs or hop extract
10 Spices or essential oils when used for flavouring purposes
11 Any substance added to food by the process of curing known as smoking
12 Carbon dioxide, nitrogen, or hydrogen when used in the packing of food in hermetically sealed containers

Table 29 *Alternative forms in which permitted preservatives may be used*

| Permitted preservative | Alternative form |
| --- | --- |
| Sulphur dioxide | Sulphurous acid or any of its sodium, potassium or calcium salts |
| Benzoic acid | Sodium potassium or calcium benzoate |
| Sodium nitrate | Potassium nitrate |
| Sodium nitrite | Potassium nitrite |
| 2-hydroxybiphenyl | Sodium biphenyl-2-yl-oxide |
| Sorbic acid | Sodium sorbate, potassium sorbate, or calcium sorbate |
| Propionic acid | Sodium propionate or calcium propionate |
| Methyl 4-hydroxy benzoate | Sodium methyl 4-hydroxy benzoate |
| Ethyl 4-hydroxy benzoate | Sodium ethyl 4-hydroxy benzoate |
| Propyl 4-hydroxy benzoate | Sodium propyl 4-hydroxy benzoate |

13   Nitrous oxide when used in the making of whipped cream

Propionic acid, sorbic acid, benzoic acid and sulphur dioxide exert their effects by virtue of their acidity. Propionic acid and its salts are permitted in bread and flour confectionery. They are valuable mould inhibitors and also prevent the development of 'rope' in these products. Sorbic acid is a useful fungistatic agent for use in flour confectionery, marzipan, cheese and other foods. Benzoic acid occurs naturally in cranberries and is added to many other foods and, in the amounts permitted, the benzoates generally are more effective against moulds and yeasts than bacteria. Benzoic acid may be added to a variety of foods to aid their preservation such as fruit juices and soft drinks, liquid coffee and drinking chocolate. Methyl or propyl 4-hydroxybenzoate may be used as an alternative to benzoic acid in some foods and in the same permitted amounts. Sulphur dioxide, in addition to being effective in inhibiting microbial growth, also helps to maintain the colour of vegetables that are going to be processed. It may be added to a large number of foods which include sausages, dehydrated vegetables, dried fruits and fruit juices.

The use of biphenyl and 2-hydroxybiphenyl in food preservation is limited as they have a strong taste and smell. They are effective in inhibiting mould growth and are used in dips to treat citrus fruits (see Table 28) and also to treat the paper used for wrapping fruit.

Nitrates and nitrites are used to aid the preservation of certain foods – being used primarily as curing salts for bacon, ham and cooked and uncooked pickled meat. Originally only nitrate was used, being converted into nitrite by bacteria present. In long curing processes today a mixture of nitrates and nitrites is used, bacterial conversion of the nitrate replacing the nitrite. But in some cases nitrite is used alone. The permitted levels are shown in Table 28. Through interaction with sodium chloride and the pH, the added nitrite is particularly important in suppressing the growth of *Clostridium botulinum* in unheated, cured meats. In canned, cured meats, which are heated in processing, the presence of nitrite in the curing process markedly reduces the amount of processing heat required to produce safe products. It is probable that the heat-damaged but viable spores are inhibited from further growth by the nitrite.

Table 29 gives the alternative forms in which permitted preservatives may be used.

### Traditional preservatives

For the purposes of the regulations most of the traditional substances such as salt or sugar which have been used for centuries are not regarded as preservatives.

The fermentation of certain foods helps in their preservation by the production of acid during the process. The fermentation of cabbage in preparing sauerkraut produces enough lactic acid to preserve the product provided oxygen is excluded. Acidity developed in this way plays a part in preserving many other foods such as pickles, cheese, fermented milk and some kinds of sausages. Another acid fermentation product is acetic acid and this is used in preserving certain meat and fish products which also undergo mild heat treatment. Diacetyl is produced by some lactobacilli and streptococci and since it can suppress the growth of some gram negative organisms it may play a role in preventing or delaying the spoilage of some dairy produce. Of course, at the same time, the growth of such producing organisms may constitute spoilage, not least because diacetyl has a smell which is associated with the smell of butter and that could be undesirable.

Salt and sugar act as preservatives by dissolving in the water of the food, forming a concentrated solution in which micro-organisms are unable to live. Water is drawn from the microbial cells into the concentrated aqueous solution and the cells become dehydrated. Salt is used in the form of brines and cures to preserve many products such as meats, while sugar acts as a preservative in jams, syrups and honey.

Wood smoke is used primarily to add flavour but it also aids preservation by impregnating the food surface with chemical substances from the smoking process. The use of a hot smoking process may in some instances reduce the number of bacteria present 100,000 fold due to the combined effect of heavy smoke and heat.

Spices are not considered to have great bacteriostatic effect in the concentrations commonly used and in fact they frequently add spores to food.

### Antibiotics

Many countries permit the use of antibiotics in food to prevent the growth of micro-organisms but their use for this purpose is limited in the UK. Some micro-organisms themselves are responsible for the production of antibiotics, *Streptococcus lactis* for example being responsible for the production of the antibiotic nisin. Nisin may be used for controlling species of *Clostridium* in cheese and also to prevent the development of heat resistant, bacterial spores which are not killed during canning and which, if they germinated, could cause spoilage such as

flat sours. Nisin is only used as an additional precaution in canned foods, the safety standards observed in canning procedures having already been obeyed. No maximum specified levels are stated for this antibiotic as it is harmless to man and has no use medically. It may also be added to clotted cream as well as to cheese and canned foods. The antibiotic nystatin may be used to prevent mould growth on bananas but it is only applied to the skin and not the flesh.

At one time chlortetracycline and oxytetracycline could legally be added to ice used to pack raw fish to prolong its life while ships were at sea. This is not now permitted. These antibiotics have also been used extensively for treating poultry in the USA, practices which may be partly responsible for the increase in tetracycline resistant strains of salmonella isolated from man and animals. Resistance can be transferred from non-pathogenic organisms to pathogenic ones, making the latter very difficult to treat. The increased use of antibiotics in animal feeding stuffs has also led to an increase in the proportion of antibiotic resistant strains of *Salmonella typhimurium* and other serotypes during the last few years. Antibiotics which are used in medicine have now been removed from the list of permitted animal growth promoting substances. Others which are only fed to animals have also been included in this ban. As a result, it is no longer lawful to supply animal feeds containing, for example, chlortetracycline, oxytetracycline or tylosin without a veterinary prescription.

### Irradiation

Although the bactericidal effect of radiation has been known for many years, its use as a means of preserving food is still very limited. The types of radiation which are of practical use in the food industry will be discussed under the following headings:

1 Ionizing radiation
2 Ultra-violet radiation
3 Microwaves

## Ionizing radiation

Ionizing radiations may be defined as those having wavelengths of 2000 Å or less, examples being X-rays, beta rays and gamma rays. They act by damaging the chromosomes present in the nucleus which are essential for cell division. This process of preservation is often referred to as *cold sterilization* as there is no heat involved. The two most widely used means of irradiating foods are gamma radiation, originating from radioactive isotopes, and electron beams, which are generated by particle accelerators. Gamma rays have much higher penetrating powers than electron beams and can therefore be used to treat packaged foods.

The unit generally accepted for measuring ionizing radiations is the rad and it is equivalent to the absorption of 100 ergs per gram of matter. The amount of irradiation that food is going to be subjected to depends on the objective, which may be to destroy all the organisms capable of multiplying in the food, or to destroy specific pathogenic organisms or to destroy a particular category of spoilage organisms. These three aims have been classified under the following headings:

### Radappertization

The ionizing radiation dose should be sufficient to reduce the number and activity of viable organisms in the foods so that few, if any (other than viruses), can be detected by the usual methods. It is equivalent to 'commercial sterility' as understood in the canning industry and should provide the same degree of safety as foods sterilized by the more conventional methods. This treatment cannot be applied satisfactorily to all foods as some undergo flavour and colour changes making them unpalatable. Its application is therefore limited to those foods which do not undergo any, or only minor, changes. Foods treated in this manner will keep for a long time, provided they are packed and stored properly and recontamination is prevented.

### Radicidation

The ionizing radiation dose should be sufficient to reduce the number of specific, spoilage, non-spore-forming pathogens (other than viruses) so that none are detected by any of the standard methods. The object is to prolong the keeping time of the food by reducing the number of spoilage organisms and to render the food safe by killing pathogens. The objectives here are similar to those in pasteurizing milk.

### Radurization

The dose of ionizing radiation in this case should be sufficient to enhance the keeping quality of the food by reducing the number of specific spoilage organisms.

The advantages of irradiation of food by ionizing radiations include the fact that it may be treated when in hermetically sealed containers and also while in the frozen state. One of the chief disadvantages is that enzymes are highly resistant to radiation and will still be capable of causing changes in the food. If the dose is increased to ensure their inactivation, the foods undergo undesirable changes in appearance, texture and flavour. As in other preservative methods the enzymes could be inactivated by initial blanching of the food before irradiation.

Foods containing high microbial loads should not be irradiated and radiation should never be a substitute for hygienic food handling. However irradiation of food for human consumption by means of ionizing radiations has not yet been sanctioned.

In addition to controlling microbial growth, irradiation will also retard the ripening of fruits and vegetables, inhibit the sprouting of potatoes and onions and also kill insects which infest grain products.

### Ultra-violet (UV) irradiation

These rays are absorbed by the intracellular DNA and RNA resulting in mutation and death.

The disadvantage of UV rays is they cannot penetrate opaque material and therefore only act on the outer surface of most foods, any micro-organisms within the food being un-affected. They have widespread applications at present which include reducing the number of organisms in the air of food rooms, treatment of growth of film yeasts on pickle or vinegar, killing of spores on sugar crystals and packaging of cheese.

The wavelength for maximum germicidal effect is 260 nm. The lethal action varies with the time of exposure and the intensity of the light – for example a lamp is 100 times more effective in killing micro-organisms at 127 mm from the object to be treated than at 2·54 m.

### Microwaves

The microwave oven utilizes high frequency radiation waves and is as safe as conventional cooking methods. These waves cook the food and can penetrate to a distance of about 115 mm but their distribution tends to be rather patchy. The food molecules are caused to vibrate and this generates heat. It is a method that can be used for cooking raw foods or for reheating foods from the refrigerated state. One great advantage of this method is the great speed at which the cooking is achieved.

The Food Hygiene Code of Practice no 9 'Hygiene in Microwave Cooking' offers the caterer advice on the use of this method.

## Controlled atmosphere packaging

Controlled atmosphere packaging (CAP) is a technique of sealing fresh foods in non-air atmospheres in disposable containers. Various combinations of gases have been tried, particularly of carbon dioxide, nitrogen and oxygen. It has been found, for example, that if fresh fish or fresh meat is stored in CAP under refrigerated conditions, the shelf life of these very perishable products is extended significantly. It is likely that CAP will be used more and more widely for a whole range of perishable products using gas atmospheres appropriate to the product. In general, it has been found for fresh foods held in atmospheres of air, carbon dioxide and oxygen at 0 to 5°C that the rate of bacterial multiplication is decreased, and the lag phase is increased when the level of carbon dioxide is increased. Gram negative spoilage organisms are more readily suppressed than gram positive ones. The process of gas storage has to be carefully investigated for the various product categories because not only the effect on the microbial population has to be considered but also possible changes in the product itself. For example, undesirable colours can develop in fresh red meat in raised levels of carbon dioxide which, in terms of selling a product, is counter-productive.

# Chapter 9

# Control of food quality

Foods are dynamic systems – they tend to change. Both growing crops and stored foods are liable to attack from animals, birds, insects and microscopic organisms, all of which cause deterioration in quality. Farmed foods – fruits, vegetables, fish and meat – contain enzymes which cause changes to the biochemical composition of the food-stuff. Initially, the changes may be advantageous, causing ripening and tenderizing, but if the changes progress too far the products may be considered to have deteriorated. Such deterioration can be judged by softness of texture, by liquefaction, or by changes in smell. At this stage the micro-organisms present tend to grow and to increase and accelerate the changes taking place. Indeed the whole chain of processes from producing and harvesting the food at the farms or catching the fish, packaging, transportation, storage processing, handling and retail sale to use in the home tends to cause damage to the foods. All along the chain contaminating organisms are added to the existing flora and there is a steady reduction in the quality of the food.

However, processing and packaging methods actually seek to maintain or improve the quality of raw food-stuffs so as to provide processed foods which are both nutritive and safe and which will keep in a good state for a target time period.

In the UK the majority of people live in cities or urban areas. This means that their food is grown and processed some distance away from where they live. The food, therefore, has to be transported to their locality, then stored for a time in the shops before it is sold and often is stored again at home before it is prepared and eaten. The manufacturer has to predict how long a time will elapse between manufacture and consumption and must ensure that the product will leave the factory in a condition which will permit it to last that long. The mode of transport and the stress of handling all have to be taken into account when predicting the product life. Care has to be taken, for example, with fruits that they are not too ripe when they are picked and that they do not get bruised in transport. Bruises are unacceptable damage in themselves but they are the source of rapid deterioration of the fruit because the enzymes in the fruit and micro-organisms take advantage of the bruised tissues.

Foods produced by fermentation processes have to be very carefully controlled in manufacture. A non-sterile food such as yoghurt, for example, has to be manufactured under strictly controlled conditions. Very clearly defined microbiological standards have to be set to account for possible microbiological changes which could occur during transport and retail storage. For example, only low levels of contaminating yeast cells can be accepted in order to ensure that the product will still be of good quality even if the yeast grows, and that signs of fermentation will not be detectable within the set storage period or by the 'sell by' date. Also, acceptable conditions of transport and shop storage have to be determined because a product will deteriorate more rapidly when kept in a warm environment.

While it is the responsibility of everyone involved in the complex chain of producing, manufacturing, transporting and cooking food to safeguard it from damage and deterioration, it is practice to give immediate responsibility for quality to specific personnel. The job of quality assurance personnel is to oversee, as far as is possible, the progress of the food from the point of production to the point of consumption and to ensure that optimum and consistent quality is maintained day by day and from one processing line to another. The quality assurance personnel supervise the chain of actions – measurement, sampling and analysis – which monitor the quality of incoming ingredients. They also monitor manufacturing processes and finished products to ensure that the correct composition is attained, the correct process is applied, the correct weight packaged and that the levels of contained micro-organisms are within acceptable limits at all stages.

Quality control personnel – who tend to be laboratory based – monitor the food quality by regular testing of food samples, equipment, packaging and environment, etc., associated with the food. By comparing the results obtained with the standards set in their records they should be able to spot changes in quality and ensure that suitable action is taken. Testing gives information about changes (such as deteriorative changes) which have already taken place. For example, how many salmonellae per gram are there already in a sample of bone meal?

Microbiological data take a minimum of one day to collect (often longer) which means that, by the time the information is available, the microbiological status of the tested item may have changed yet again. How, therefore, can such data be used?

The answer is that it can be used to *predict* quality. If the cell counts of certain spoilage bacteria are outside the set limits, it is likely that the batch of food from which the sample was taken will not keep as long as its target life. The presence of certain microbial species may indicate the possibility of that batch of food causing illness in consumers. Equally, micro-

biological data may throw light on past events – for example, through the presence of certain bacterial types it may be revealed that a manufacturing process has been inadequate. An example of this would be the presence of indicator organisms, such as coliforms, which could well indicate that a failure in hygiene had occurred. The events have already happened but, with that knowledge, suitable corrective or protective action can be taken to restore the manufacturing process so that the quality of the product is returned to the desired level.

Quality assurance and control procedures aim to maintain or improve quality and prevent contamination and deterioration occurring. To achieve this, the individual standards of the manufacturing team are collected together so that all the particular criteria of the team are covered. The overall quality criteria for a product, therefore, include consideration of the cost of the components, the sales image of the product, its nutritional quality, its keepability, its texture, its freedom from pathogenic and spoilage organisms, etc.

But food manufacturers are also required to comply with standards imposed on them by national legislation and by the legislature of the

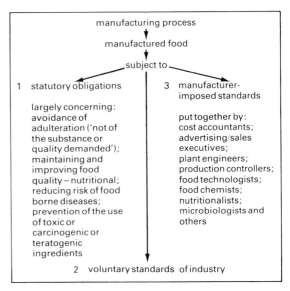

Figure 67 *Pressures for quality control*

countries to which the food is exported. Generally, legislation is aimed at improving food quality and preventing adulteration and unfair practice. (See Figure 67).

## Statutory obligations

The law in the UK requires that 'food must be of the substance and quality demanded by the consumer'. In practice, this means that if a food contains foreign objects, such as glass, plastic, fur, rodent droppings, etc., or contains adulterating substances, or is mouldy or 'off', or is proven to have caused food poisoning or food borne disease, legal proceedings may be brought against those deemed to be responsible. Liability may be attributed to retailers, food handlers, manufacturers or importers – with the object of preventing practices which encourage the sale of unsafe foods. To avoid this risk, strict control is necessary throughout the production, transport, wholesale and retail chain. Cases arise where dairies and food manufacturers are taken to court over offences under the Food and Drugs Act, 1955. Since such cases are usually heard in public it is relatively easy to find out about the failures of manufacturers to meet the statutory obligations but, for obvious reasons, internal problems in meeting self-imposed standards are generally not publicized.

The Sale of Food and Drugs Act, 1875 was the first important legislation. Today, The Food and Drugs Act, 1955 and the statutory instruments made under it embody the current UK legislation (see Appendix 1). This legislation has helped to prevent a ruthless approach by manufacturers who at one time might have been tempted to cut corners to increase profitability; it also protects the health of consumers and seeks to ensure fair practices. On an international scale, co-ordination of legislation and objectives is debated through the FAO/WHO Food Standards programme and implemented by the Codex Alimentarius Commission.

But today, even after more than 100 years of food law, there are very few microbiological standards which have to be met by law. This is because for legal purposes a test must be infallible and it has been shown that microbiological counts cannot be totally reliable. However, there are many compositional and labelling standards which are controlled by law.

In the UK milk quality is closely controlled. Pasteurized milk has to pass a chemical test for the absence of the enzyme phosphatase which is destroyed by correct pasteurization procedures. Although pasteurization aims to reduce the number of bacteria in milk and particularly to destroy those which cause disease, no bacteriological test is required.

There is a reluctance in the UK to impose statutory microbiological standards for foods when it is known that even when using standardized methods of analysis, widely differing results can be obtained on the samples analysed at different laboratories.

## Industry infrastructure

Another aspect of food quality control is demonstrated by the co-ordinated activities of the many people or groups and companies working within a single industry. Two examples of this, the first showing how good quality pasteurized milk is produced and distributed and the second how chilled, fresh foods are distributed, illustrate the point.

### Pasteurized milk in the UK

Because milk is so much a part of the staple diet of people, both because it is nutritious and because as a nation we can be self-sufficient in it, the Milk Marketing Board (MMB) was set up. The Board controls the marketing and quality of this valuable food commodity. Figure 68 summarizes how the overall good quality of milk is maintained. Failure by producers or processors to meet the standards set penalizes them financially and the poor quality milk is redirected to a more suitable use.

### Chilled foods

Foods are protected by a voluntary code of practice which is gaining acceptance within the membership of the trade. Fresh products must be chilled all the way along the chain from

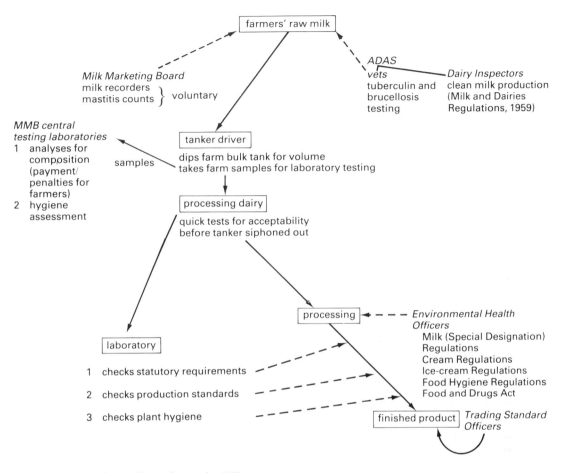

Figure 68  *Control of milk quality in the UK*

production to retail sale. Failure to do so means that micro-organisms in the food together with the food's own enzymes and possible attack by insects will cause a more rapid deterioration of the food. Refrigerated lorries are designed to maintain a low temperature, close to 1°C, and not to cool hot food. Good chilling at the point of production is essential *before* loading. The chilled products should be unloaded without the doors of the refrigerated lorry being opened so frequently that the interior warms up significantly. The food should be off-loaded quickly so that its chilled temperature does not rise and it should be stored in chilled store rooms straightaway. Display cabinets are very rarely ideal in their design since sunlight, electric light bulbs, the amount of space open and accessible to the

shopper, draughts and the levels of loading all affect the temperature the cabinet can maintain. Ideally, throughout all these operations, the food temperature should not rise above 2°C. It is obvious that considerable co-operation between the many companies of food producers, distributors and retailers is necessary if the quality of the food is not to suffer as a result of protracted periods of raised temperature.

## Manufacturer-imposed conditions

Manufacturers set themselves certain standards to attain. They do this because they know which quality and type of product sells best and it is on the success of their products that their business thrives or fails. Thus manufacturers will try to control the quality of the raw products they buy and will strive to produce a product of consistent quality. They can only do this by keeping a very close eye on every stage of the process.

The manufacturers will decide on the degree of control they wish to exercise or can exercise on their suppliers at their factories. They will certainly examine the raw materials on reception as they will not wish to pay for goods which are not up to the required standard. The measure of control and testing exercised in the factory is in the hands of the manufacturer who should therefore send the products out with a good knowledge of their average quality. The manufacturer may have some influence (varying from a little to a lot) over the conditions of distribution and may also exercise some control at the point of sale. Control comes as a result of examining the product, looking at it, testing it, measuring some of its qualities, such as its temperature, viscosity, acidity, and by monitoring the way the processing machinery functions, the way it is cleaned and so on. An awareness then develops of what is the correct situation and changes from the normal which warn of problems ahead can be readily detected. The monitoring of the production has to be carried out using a system which is accepted by all the people who have to operate it. It is no use monitoring a large number of processes and obtaining many test results only to file the results away and take no further notice of them. In a food manufacturing process there are several different groups of people whose joint agreement is required before the high standards which develop as a result of monitoring the process can be achieved (see Figure 67).

One aspect of control involves monitoring the costs of ingredients, composition, plant wear and tear, investment programmes, power, water and effluent. The cost to a manufacturer of producing a food product derives from the cost of its raw materials and the cost of manufacturing. The profit which can be made on foods depends on minimizing processing costs but still selling a good product of reliable quality. It is necessary, therefore, to check the composition of incoming ingredients to ensure that those which do not come up to standard are rejected. Control of physical processes and basic ingredients should mean that manufacturers make the good quality product they expect to make and do not have to reject batches of sub-standard product: for example, dough which is made that rises satisfactorily to make good bread, yoghurt that coagulates within the set time limits or cheese that is not adversely affected by bacteriophage. Failure to meet their own criteria can only mean financial loss to manufacturers.

Sales and advertising departments are concerned with the qualities of the food and its packaging and presentation, all of which will aid its sales. These aspects are largely determined by market forces.

The physical properties of the food, such as flow, viscosity, density, mixability, homogeneity and heterogeneity of its thermodynamic properties, all have to monitored to ensure that the processing produces the expected product. Unnecessary losses of product and wear and tear on the processing equipment are also to be avoided.

The chemistry of both ingredients and product – nutritional, compositional, pH, additives, emulsifiers, stabilizers, colours, flavours, contaminants (metal and other) – all have to be closely monitored through measurement and testing to ensure that every batch of product is of the quality desired.

## Microbiological quality control

Microbiological quality control covers the testing procedures which ensure that neither food borne disease nor food poisoning will result from consumption of the product. Microbiological quality control is also concerned with monitoring ingredients and product for levels and types of food spoilage organisms and in part contributes to the data required for the setting of the 'sell by' date. This gives the length of time that a product should keep in good condition and be of stable quality after manufacture.

Microbiological tests are, of course, integral when products are produced through the action of bacteria or other micro-organisms as are, for example, yoghurt, cheese, beer and wine.

In determining the microbiological status of a food or piece of equipment, two factors have to be considered:

1  What is the level of reliability of a single microbiological test? Are several similar tests required and, if so, how are the collective results put together?
2  What is the sample taken and analysed representative of? How can the general quality of a number of packages or boxes of food be judged? Do the characteristics of the sample represent those of the whole batch and/or do they represent those of only an area perhaps where deterioration has started. Has a problem (old or new) been identified or missed?

Microbiological analyses involve many steps and different techniques to find out different things. It is not the place here to go into the details of those techniques but, as an example, the following list outlines the steps involved in analysis of the number of living organisms present:

1  Take the sample.
2  Put it into a container.
3  Transport it to the laboratory (this may take a minute or two or several hours).
4  Take a subsample out.
5  Mix it with a diluting fluid.
6  Make several dilutions.
7  Culture the organisms in the diluted sample.
8  Look at the results obtained.

Now suppose some of the organisms which were alive in the original food die in the time between taking the sample and analysing it; suppose others increase in number; some techniques used in the laboratory might kill organisms or might only show up certain species present, leaving others present undetected. One colony on a plate may result from a single cell or aggregation deposited there. A good micro-biologist is aware that these and other possible factors will influence the results recorded. Of course proper techniques reduce some of these problems; while other problems may be recognized but unavoidable.

If the same technique is performed regularly so that records of results on many samples are obtained, then, when a variation in results is obtained, the microbiologist must ask herself or himself – is the variation due to changes (possible significant changes) in the sample or is it due to the technique used? Does the latter vary every time the same test is carried out? Also, when a result is obtained, what does it mean?

A numerical result simply means that, by the technique used on that occasion, a certain number of organisms were detected. This is not absolute because allowance has to be made for those not detected and for the possible increases that may have taken place before analysis but after sampling. From a single test, the number of organisms detected cannot be taken too literally. The number is really just a guide as to the amount of organisms present. This is because using the common counting techniques, even in the best laboratories, there are so many areas of inaccuracy that little importance can be placed on results which show a difference between say 1.2 and $2.5 \times 10^6$ organisms per millilitre. Obviously, the more refined and controlled a technique is, the more accurately the microbiologist can determine true differences in results.

If the same technique, as near as possible, is used by one laboratory only, the results obtained can be very useful. Although they cannot be thought of as absolute, they provide useful comparative data from one sample to the next. When analyses are performed by several laboratories, an extremely careful standardization of techniques is required to permit comparison of results. This is why on the few occasions when microbiological counts are required by law (for food standards testing purposes) both the techniques and the equipment to be used are very closely defined.

So, using a certain set of techniques, a sample is deemed to contain a certain number of organisms of a particular species. The data are recorded and a comparison can then be made with previous records held. Is the count, and therefore the sample, within or outside acceptable limits?

As mentioned above, a common microbiological test is to count the number of living organisms per gram or per millilitre; and these may be identified in some way as a group or as a particular species. For example: $1.2 \times 10^5$ organisms/gm, or $3.2 \times 10^3$ coliforms/ml, or total viable count $1.0 \times 10^6$/ml.

A gram or a millilitre is a very small quantity. The implication behind these figures is that the organism count recorded is representative of the sample from which it came. Perhaps a 500 gm sample arrived at the laboratory and the microbiologist took as much as 50 gm from it and tested that, recording the results found per gram. How representative of the 500 gm was the 50 gm? How representative of the bulk quantity (1 box out of 1000 for example) was that 500 gm sample? If two samples had been taken from different boxes instead of the one 500 gm sample and very similar results had been obtained, perhaps more reliance could be placed on them as being representative of the mass of boxes. But suppose different results had been obtained. What would it mean? What if the results were very or fairly similar? How are they to be interpreted – is the batch of food acceptable or not? Can the testing data be extrapolated in relation to a batch of food bigger than the sample? If it cannot, is there any point in doing such a count? So the judgement which has to be made rests on the fact that the sample is assumed to be representative of a larger quantity and what is the chance of that being true? It is totally dependent both on how the sample was taken, which in turn is dependent on a good knowledge of the microbiological and physical properties of the food, and on the application of proper statistical sampling techniques.

Liquids can be easier than solids to sample because they flow and can therefore be well mixed. A tank of milk, containing perhaps 10,000 litres, can be mixed for about 2 minutes with the tank's agitator and if several samples are taken and carefully analysed, counts showing the same number of organisms will be obtained. Conversely, samples from bulk milk which has been standing a while and has not been mixed will show considerable variation in counts and reliable interpretation will be difficult.

If the quality of piped water has to be tested, care has to be taken that organisms which are lodged in the intricacies of the tap mechanism are not included in the samples. These organisms must be flushed away and the tap sterilized before sampling, otherwise the results would be meaningless.

Solid foods, like chicken carcasses, are difficult to test. Research has shown that different areas on a chicken's carcass have very different bacterial loads. For example, high skin counts are found under the wings and near the anus and cloacal region; lower counts occur on the breast skin. All skin counts are very much higher than those of the flesh. A chicken might weigh 1500 gm. If a small subsample of 10 gm were taken to be analysed, should the results then calculated as a count per gram, be taken to refer to the whole bird? Would the data obtained relate to other chicken carcasses and be representative of them?

Obviously any animal carcass shows a similar irregular distribution of micro-organisms. The way sampling is tackled, in the face of such irregularities, is to identify which problem is being examined and therefore to what the data relate. Then the appropriate sample from the whole macerated carcass (or several carcasses mixed), or from the skin only, or from a selected area of the skin, can be taken. But not only should it be realized that the micro-organisms may be irregularly distributed throughout a single food item, but they may also be so throughout a mass such as a quantity of minced beef. When such a mass is canned, bottled, or put into cartons, some units will contain a higher number of organisms than others.

As another example, suppose a yoghurt-filling machine was not cleaned properly and was instead allowed to stand, uncleaned, overnight so that yeasts grew in the yoghurt residues. Then the next day it was used to fill 10,000 cartons of yoghurt. The first cartons to be filled would be badly infected with the poor quality yoghurt left in the machine, while later cartons would tend to be less affected and contain yoghurt more typical of the normal product. It is known that some of the 10,000 cartons are much inferior to the rest. But if only one carton (or perhaps five cartons) were taken for microbiological analysis, what would the results indicate? Which cartons should be taken and how many?

Sampling of foods using a statistical sampling plan is of great assistance in resolving the problems posed by a situation such as the one described. In statistical sampling, samples are taken on bases which take into account the quantity of food to be judged, how the food is apportioned (that is, the size of the container), the type of food, the tests to be undertaken and the types and numbers of organisms to be looked for. The results of the subsamples analysed are then recorded. Statistical sampling also embodies a statistical approach to the results obtained for a batch of food tested, whether it is the production of a shift at a factory, the production of a single machine per day, a pallet of tins of food or whatever. Within the results of a single batch, a certain level of failure will be allowed as acceptable, that is, a certain maximum number of results may fall outside the microbiological standard set. If fewer results fail than allowed in the sampling plan, then the batch of food tested is acceptable. If more results fail than allowed, the batch fails and a decision has to be taken as to what to do with the food. It may be reprocessed, it may be directed to a use different from that intended; in the case of supplies the food may not be purchased or a lower price may be negotiated.

With process equipment a different approach may be used. Certain areas of the plant, such as the sides of a tank, particular bends, fillers, taps, pumps, screw threads etc., will be sampled regularly by standardized techniques to provide data – such as the microbiological condition after cleaning in place (CIP). From these data the acceptability of the cleaning process can be determined. In addition, other areas of the plant will randomly be sampled to check on the assumptions that are made as a result of the regular tests.

But how is a microbiological standard set in the first place? This is not easy to answer. Broadly, experience built up by the food manufacturer or by public health authorities indicates what the characteristics of the food are, which spoilage organisms and pathogens it may carry, what problems are likely to occur and how the processing should deal with them. Using this knowledge, manufacturers, producers and importers can then set their own internal standards both for the supplies they buy and for the products they produce, though they do not often make these standards public. They may base these on their own experience or on recommendations of respected bodies, such as the International Commission on Microbiological Specifications for Foods. Food legislation, as already mentioned, is not concerned directly with microbiological standards. However there are standards which have to be met for milk (see Appendix 1) and, under EEC legislation, for bottled spa water. It is more common for the law to require that food is treated in such a way that problem organisms such as salmonellae are eliminated and proof must be given that the treatment given has been successful. The best examples are the phosphatase and $\alpha$-amylase tests for pasteurized milk and liquid egg respectively, though these are chemical and not microbiological standards.

However now that statistical sampling is more firmly established as a reliable method of testing, it does seem likely that more direct microbiological standards will soon be required by law.

To maintain a constant food quality, therefore, a manufacturer has to closely monitor the entire process of manufacture. The quality is maintained both by physical checks, such as measurement of temperature and pH, and by taking samples for analysis. The types of analysis include tests for the presence or absence of micro-organisms. Because of the irregular distribution of organisms in batches of foods, statistical sampling* is the best approach. Good quality food leaving a point of production may deteriorate rapidly in improper distribution conditions, but co-ordinated effort within an industry, such as the chilled food industry, will help to avoid that happening. Legally, certain standards are imposed to protect the public.

'Sell-by' or 'use-by' dates are now often attached to the packaging of perishable food stuffs. These are dates by which the manufacturer judges that the product will be close to the end of its peak condition – both in terms of microbiological and chemical conditions.

## Microbiological techniques: an outline

The techniques available to the microbiologist can be divided into categories:

1   Techniques involving direct observation.
2   Techniques involving counting the organisms – these methods require taking samples and counting either the total number of organisms (living and dead), the total number of living organisms or estimating the number of living organisms.
3   Analyses for specific organisms but not necessarily concerned with their absolute numbers – looking for indicator organisms.
4   Using the activities of the micro-organisms to indicate their presence and/or their numbers – these include observing pH changes, production of gas and reduction of dyes and looking for direct evidence of biological change such as evidence of spoilage.
5   Analyses for the presence of microbial products such as toxins.
6   Undertaking tests which involve exposing a microbial population to certain environmental circumstances and observing the reaction.

---

* See *Micro-organisms in foods*, Volume 2. 'Sampling for microbiological analysis: Principles and specific applications', The International Commission on Microbiological Specifications for Foods (University of Toronto Press 1974).

### Direct observation

With the naked eye certain changes in the texture and the colour of a food, due to microbial activity, can be observed. The production of surface slime on meat, coagulation in souring milk, mould on spoiling fruit and other factors, such as the presence of gas, are all visible signs of microbial activity.

Small samples of liquid can be examined under a microscope, possibly using staining techniques, to ascertain the variety and morphology of the species present. When cultures have been set up in the laboratory, observation of their colonial morphology, pigmentation, colour, response to the growth medium and other features can help in identifying the organisms present.

### Counting organisms

The questions 'How many?' and 'Which types?' are very often foremost in the microbiologist's mind since such information is extremely valuable in assessing the nature and extent of a problem. It is sometimes useful to calculate the number of all the cells, living and dead, present per gram or per millilitre. This provides a quick assessment of quality without having to wait while cultures are incubated. Generally the total count involves the examination under a microscope of a thin film of sample. The count per unit volume can be calculated from knowing the area and the depth of film viewed.

Total viable count techniques are often used and depend on individual viable cells or cell clumps increasing and producing visible colonies on a solid nutrient medium. From knowing the volume of sample added to a plate, the number of colonies counted per plate can be computed to give a count per unit quantity analysed. A total viable colony count indicates the microbial load which potentially can increase and, from knowing this, an assessment can be made of the degree of hazard associated with the presence of the organisms. Regular counts are very useful for monitoring quality.

'Most probable number' techniques involve the inoculation of a series of growth media with a range of sizes of sample. If a sample contains

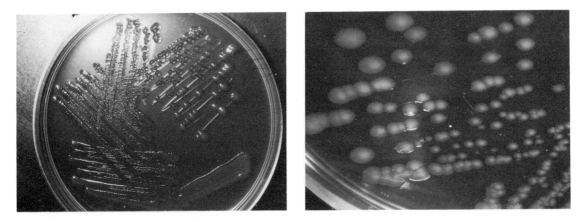

Figure 69   *Isolated bacterial colonies growing on solid media in petri dishes*

the organisms looked for, growth will follow and a positive test will be recorded. Conversely if a sample does not contain the organisms, no observable change in the medium will occur. From the combination of positive and negative results the most probable number of organisms in the original sample can be estimated. This technique is covered in Chapter 11.

### Analyses for specific organisms

To analyse for the presence of specific organisms, certain techniques have to be applied and carefully used. Perhaps the most important of these are manipulating the composition of the medium and exploiting the different growth characteristics of the organisms in order to favour one group or species and to discourage others. Once individual genera or species are isolated, then other important techniques which exploit subtle biochemical differences or detect serological differences can be used for more specific identification (see Figure 69).

A medium must provide the minimum growth requirements of the organisms of interest – that is, proteins and amino-acids, carbohydrates, vitamins, minerals and water, as necessary. Some organisms are very fastidious in their growth requirements (streptococci have very specific growth requirements for vitamins for example); others are less demanding and are able to synthesize their requirements from the restricted nutrients available. Media which supply very much more than minimum growth requirements will permit more rapid growth (see Figure 33, page 34). Careful choice of the medium used determines which organisms in the mixed population contained in a sample will be able to grow. No one medium is able to support all types of micro-organisms, so the choice of medium is the first stage in isolating particular species, though at the same time it must be remembered that other organisms present in the sample will not be shown up using that medium. Often at the first stage of sample analysis several subsamples are incubated on different media, each of which is designed to favour certain types of organism. In this way the range of organisms present may be more fully appreciated.

'Non-selective' growth media are those which support the growth of a wide range of organisms. Tryptone soya broth is an example and nutrient broth is another (see Table 30) and both of these are commonly used in solid form to perform total viable bacterial counts.

Media are either liquid or solid – each has its uses. Liquid media permit a whole range of organisms to grow together and the changes in the medium as the organisms grow will be due to the combination of the contributions they all make. The organisms will not all grow at the same rate – and the growth of some and the changes produced as a result will encourage or suppress the growth of others. The numbers of

Table 30    *Examples of the composition of media*

1    *Tryptone soya broth\**    g/l

| | | |
|---|---|---|
| Tryptone | 17.0 | protein/ |
| Soya peptone | 3.0 | amino acid |
| Sodium chloride | 5.0 | |
| Dibasic potassium phos-<br>phate | 2.5 | |
| D-glucose | 2.5 – carbohydrate | |

pH 7.3 approximately
(near neutral)

*Comment*: a very nutritious medium supporting growth of a wide variety of bacteria and some fungi, including many fastidious organisms.

2    *Nutrient broth†*

| | | |
|---|---|---|
| 'Lab-Lemco' powder | 1.0 – a meat extract (pro-<br>tein and amino-acid | |
| Yeast extract powder | 2.0 ⎫ | protein/amino- |
| Peptone | 5.0 ⎬ | acid/vitamins |
| Sodium chloride | 5.0 | |

pH 7.4 approximately

*Comment*: supports growth of many non-exacting organisms.

3    *MacConkey's agar†*

| | |
|---|---|
| Peptone | 20.0 – proteins/amino acids |
| Lactose | 10.0 – carbohydrate fermented by some species to acid |
| Bile salts | 1.5 – tolerated by enteric organisms |
| Sodium chloride | 5.0 |
| Neutral red‡ | 0.05 – pH indicator |
| Crystal violet | 0.001 – inhibits gram positive bile tolerant cocci |
| Agar | 15.0 – solidifies medium |

pH 7.2 approximately

*Comment*: Colonies of lactose fermenting organisms appear as very red colonies (for example, coliforms, *E. coli*); streptococci are inhibited; non-lactose fermenters are very pale (for example, salmonellae, shigellae).

4    *Baird-Parker's medium†*

| | | |
|---|---|---|
| Tryptone | 10.0 ⎫ | |
| Lab-Lemco meat extract | 5.0 ⎬ | nutrients |
| Yeast extract | 1.0 ⎭ | |
| Sodium pyruvate | 10.0 – selective growth stimulant | |
| Glycine | 12.0 | |
| Lithium chloride | 5.0 | |
| Agar | 20.0 | |

pH 6.8

| | | |
|---|---|---|
| Egg yolk emulsion | 50 ml | diagnostic agents |
| 1% w/v potassium<br>tellurite solution | 11 ml | added to<br>the melted<br>sterile medium |

*Comment*: after 24 hours 37°C incubation, colonies of *Staphylococcus aureus* on this opaque medium are shiny black with a clear zone around them in which may later develop an opaque zone. No other organisms have this appearance on this medium.

5    *Wort agar*

| | |
|---|---|
| Malt extract | 15.0 |
| Peptone | 0.78 |
| Maltose | 12.75 |
| Dextrin | 2.75 |
| Glycerol | 2.35 |
| Dipotassium hydrogen<br>phosphate | 1.0 |
| Ammonium chloride | 1.0 |
| Agar | 20.0 |

pH 4.8 approximately

*Comment*: a medium for the growth of yeasts and moulds – the pH is close to their optima and inhibitory to most bacteria.

*Standard methods for the examination of dairy products*, 11th edition, (American Public Health Association 1960).

†Nutrient broth – Oxoid formula CM1; MacConkey's agar – Oxoid formula CM109; Baird-Parker's medium – Oxoid formula CM275 based on A. C. Baird-Parker, *J. Appl. Bact.* 25(1) 12–19 (1962).

‡Other indicators – such as bromo-cresol purple may also be used.

the different organisms present will vary according to the temperature and time of incubation.

If the media used are solidified by the addition of the gelling agent agar, individual cells, or clumps of cells, derived from the sample can grow into colonies which are physically restrained from mixing with the other cells and colonies by the agar. After incubation, the different species present will be partially identifiable by their different colonies. Pure cultures can be made through the culture of cells taken from well isolated individual colonies.

If selective agents are incorporated in the medium used, the growth of many species which would otherwise grow can be suppressed, while allowing the growth of the species which are of interest. Antibiotics and certain anti-microbial agents are commonly used for this purpose. High concentrations of sodium chloride will favour halotolerant or halophilic species while suppressing others. Bile salts are tolerated by enteric organisms but not others. Further selection can be achieved by manipulation of the pH. Acids may be added to lower the pH and buffers to stabilize it. But where inhibitory chemicals are used they may be mildly toxic even to those species which tolerate them and there is the possibility of injury, damage and perhaps destruction of the species sought, which must be remembered when the results are assessed.

Then if indicator components are added, particular identifying growth characteristics will be emphasized. pH indicators are commonly used to show, by distinct colour changes, when pH in the medium has significantly altered. Individual colonies (when grown on solid media) can then be distinguished from others. Mac-Conkey's agar medium is a good example of a medium which contains both inhibitors (selective agents) and indicator components (see Table 30). *Staphylococcus aureus* can produce distinctive black colonies on Baird-Parker's medium. Since these organisms can produce enterotoxin (and thus food poisoning) they would be highly undesirable organisms in cooked custard pies or other cream products. A manufacturer making such items should check regularly that the products conformed to a set standard with regard to the presence of staphylococci, and a quick sure method of identification of staphylococci would be needed for routine sample analysis.

Another selective procedure used is to mildly heat a sample in order to kill the heat sensitive cells and leave behind the thermoduric and very heat resistant cells. For example, the number of heat resistant cells in milk could be assessed by heating it at 63.5°C for 35 minutes and subsequently culturing and counting the remaining cells. Heating more severely – at 70 to 80°C for 10 to 15 minutes – would destroy all vegetative cells leaving only bacterial spores which could then be cultured and counted. However, the remaining organisms after heat treatment might be heat damaged and the initial treatment of the samples would have to take this into consideration if a true reflection of the number of viable organisms remaining were to be obtained.

Then, by exploitation of the different temperature preferences of organisms, further distinctions can be made. Incubation at 10 to 15°C favours psychrotrophic bacteria, yeasts and moulds; 37°C favours the mesophilic bacteria and higher temperatures favour the thermophiles. Moulds and yeasts can often be separated from bacteria by the use of nutrient media adjusted to a pH of 5 or below and then incubated at 25°C, since most mesophilic bacteria of importance in food will then be suppressed.

Analyses for specific organisms depend on a thoughtful approach to the procedures used. The population in some samples may need to be thinned, in others concentrated. In the former case the sample can be accurately diluted in a diluent solution before it is cultured; in the latter case, centrifugation and/or membrane filtration (see page 170) or enrichment may be applied. Salmonellae are pathogens and their presence in foods even at very low levels is undesirable. The actual acceptable level depends on the food type, but with baby foods a very low level indeed would be demanded since babies are more liable to succumb to enteric illness than adults. Where salmonellae are looked for, enrichment is often

sample
(very low numbers of
salmonellae per gram) | Salmonellae could be missed if small samples (1 gm) were taken

↓ 25 gm

1   non-selective
enrichment broth
37° for 24 hours | Salmonellae increase to detectable numbers. Non-salmonellae increase too

↓ 10 ml

2   liquid selective
enrichment
37° for up to 72 hours | The selective agent (for example tetrathionate) inhibits the growth of many organisms allowing some, including salmonellae, to grow

↓ loopfuls streaked on to

3   solid selective media
37° for 24 hours | Salmonellae show distinctive colonies after incubation. On Taylor's xylose lysine desoxycholate agar, salmonellae colonies are red with black centres, shigellae are uniformly red; on brilliant green agar salmonellae are pink or red surrounded by a zone of bright red medium. Lactose fermenters are inhibited

↓ selected colonies picked off

4   identifying tests | Salmonellae are precisely identified by a series of further cultural tests, biochemical tests and serological tests

Figure 70   *Principles of analysis for salmonellae*

the technique required – to allow the few salmonellae present the opportunity to increase to a detectable level. The growth of all organisms present in the sample is permitted by the use of a non-selective medium. Then selective procedures are used to isolate salmonellae from the total population followed by positive identification (see Figure 70).

Identification can involve several stages:

The *appearance of colonies* on non-selective and selective agars in the isolation procedures already described means that pure cultures can be produced simply by picking off typical well isolated colonies and inoculating them into suitable solid or liquid media.

*Biochemical tests* for the characterization and positive identification of pure cultures can involve a series of tests to detect the significant identifying metabolic activities of the organism. The tests are for reactions involving protein, amino-acids, proteolytic activity, carbohydrate metabolism, fat metabolism and other reactions. Such reactions may be detected by a colour change (due to production of $H_2S$, pH indicator colour change or dye reaction), by development of opacity, by production of gas, by a change in texture (owing to liquefaction of gelatin or coagulation or digestion of milk protein) by hydrolysis of starch, etc. The particular combination of positive and negative results should permit an identification. The tendency today is towards computer analysis of the results when multiple tests have to be set up. This avoids placing too much emphasis on the result of any one test in the set of tests and is particularly valuable when variants are found.

*Serological tests* may be required when species have been identified – as is often the case with salmonellae, streptococci and other organisms.

Pure cultures of cells are grown in liquid media for many biochemical identification tests, using cells isolated from the sample being investigated. Liquid media can also be used for purposes of resuscitation at the first stages of isolation of species when samples may be suspected to contain injured cells.

Solid media, on the other hand, may often be used as the first stage of culture, because, although mixed cultures will be produced, the isolated colonies will permit a certain degree of identification and estimation of the microbiological condition of the sample. Further use of solid media will allow purification of cultures and study of the reactions of the organisms on various media will lead to further identification. Solid media are also used for plate counts.

*Types of sample taken for culturing techniques*

Water and solid and liquid foods are usually subsampled and data are recorded as counts per gram or per millilitre.

Obviously, processing equipment cannot be sampled in the same way – and so, instead, contamination levels of standardized surface areas are recorded. To obtain this information, the organisms present have to be removed from the surface. The techniques to do this involve swabbing or rinsing a minimum surface area of 100 cm$^2$ or a specific area of piping. The organisms are detached from the swab by swirling it in sterile liquid, or are already contained in the rinse solution and are then cultured and counted. Alternatively, surfaces may be examined using contact plates. These are prepared plates of culture medium of known area, which are directly impressed on to the test surface. The organisms adhere to the gel and the plate is removed and incubated to permit their growth to form visible colonies which are then counted. When air is sampled, a volume of air, measured in litres, is assessed for its microbial load. Air is drawn into a sampler and the organisms present are collected on a membrane, or directly on to a strip of nutrient medium, which is then cultured.

*Effect of injury to micro-organisms on cell counts*
It has already been mentioned that injury is sustained by cells when they are heated – as is the case, for example, in food processing – and when selective agents are used in culture media. For example, staphylococci are quite tolerant to sodium chloride – to a greater extent than most other non-halophilic organisms. However, if sodium chloride at 7 to 10 per cent w/v is added to normal growth media, heat injured staphylococci cannot grow and uninjured staphylococci can. This suppression of growth could lead to the false conclusion that staphylococci were not present, or were only present in much lower numbers than was in fact the case. Because it is realized that injury occurs, it is now becoming widespread practice to allow injured cells present in a test sample a period to recover from their injury by incubating the sample in the non-inhibitory medium for an hour or two. The time period allowed is insufficient to allow increases in cell numbers. The sample containing the recovered cells is then subjected to the normal analytical procedures.

*Microbial activities*
Sometimes it is more convenient and quicker to assess the microbiological quality of a food sample through observing effects due to biochemical activities. The two main tests used are visible changes due to pH and visible changes due to reduction of added dyes.

*pH change*
In many media one of the components is a pH indicator which changes colour when the pH changes. Bromo-cresol purple is purple above pH 6.8, but when the pH drops, it changes and at pH 5.2 it is yellow. When coliforms grow in MacConkey's broth which contains bromocresol, the acid they produce causes the pH to drop and the medium changes from purple to yellow. Other commonly used indicators are:

| | |
|---|---|
| Phenol red | range pH 6.8–8.4 change from yellow to purple pink |
| Bromothymol blue | range pH 6.0–7.6 change from yellow to blue |
| Methyl red | range pH 4.4–6.0 change from red to yellow |

*Dye reduction*
Another convenient way of estimating the microbial load of a food sample is to use a dye reduction test. This depends on an alteration in the reduction-oxidation potential of a medium which is brought about by the activity of the micro-organisms present and detected by a colour change. In principle, the activity of the organisms present is estimated by adding the dye to the food sample and incubating the sample at an appropriate temperature – the whole test is carried out under carefully standardized conditions. The dyes which are commonly used are methylene blue and resazurin, although there are many others available.

A methylene blue test* is used to assess the hygienic quality of raw and pasteurized milk. It

---

* A specific and detailed test procedure is laid down in The Milk (Special Designations) Regulations 1963.

involves the addition of 1 ml of methylene blue to 10 ml milk. The milk is then incubated at 37°C for 30 minutes initially and examined at regular intervals for decolourization to white. The more active the organisms present, the more rapidly the methylene blue is decolourized and when the test is performed under carefully standardized conditions, using appropriate controls, it can be used to indicate the potential keeping quality of the milk or to grade it. Methylene blue tests are also applied in the assessment of the hygienic quality of cream and ice-cream.

Another dye reduction test also used with milk is the resazurin test. As resazurin is reduced it changes from blue/mauve to pink to colourless. Since the colour change is gradual, in testing procedures the actual colour obtained is compared with standardized coloured glass discs in a Lovibond (Tintometer Sales Ltd) comparator, and given a value from 0 (colourless) through 1 (pink) to 6 (blue). Resazurin tests can be applied to milk in several ways – from a quick 10 minute test at 37°C to detect unsatisfactory raw milk to a 2.5 hour test at 37°C for hygienic quality. The way in which such tests are performed is very specific and carefully controlled.

The dye triphenyltetrazolium chloride (TTC) is colourless when oxidized and red when reduced. One example of its use is as a convenient demonstrator of the irregular distribution of active organisms on the white skin on a chicken carcass. The test is carried out simply by applying a thin layer of the dye to the skin. Red areas appear at sites of high microbial concentration. This dye is also used in a test to check whether antibiotics are present in milk. If an active culture of antibiotic sensitive organisms are added to milk containing antibiotic, the culture will not be able to grow and the failure to grow can be indicated by the failure of added TTC to turn red because the oxidation-reduction potential of the milk remains the same and the dye remains in its oxidized colourless form.

### Analyses for microbial products

The analyses for toxins – such as staphylococcal enterotoxin, botulinum toxins and aflatoxins – are generally complex biochemical procedures. In the case of aflatoxin risk, it is certainly better to assess the concentration of aflatoxin since it is that which affects the quality of the food, and, other than identifying the mould species found, not to count the mould concentration. It is a dubious procedure to 'count' moulds where they occur as hyphal particles because the lengths of the hyphal pieces can be so variable and grossly affected by the sampling and analysing techniques.

### Tests involving a challenge to the microbial population

With mass-produced food it is often the practice to withdraw a number of samples from each production run and to use these for quality assessment purposes. Some of the samples will be put into challenge storage – one batch at a temperature close to the expected storage temperature and another batch at a higher temperature – and all samples will be kept for a pre-set period. At the end of that period the samples will be examined for spoilage by residual organisms within the food – possibly only by observing the external appearance (swollen containers due to gas production) or by opening the container and examining the contents.

A different approach is to challenge products by the experimental addition of microorganisms. Intermediate moisture foods such as cakes might be challenged with certain moulds to ascertain safe storage conditions for good keeping quality; sugar syrups might be challenged with yeasts, etc. The reaction of the challenging organism provides useful information towards product design and/or storage environment requirements.

When chemical sterilants are used, it is important that their anti-microbial activity is appropriate to the purpose. *Suspension tests* permit the correct dilution of sterilant and exposure time to be calculated. The tests involve the incubation of the sterilant at an experimental dilution with a culture of test organisms. Sub-samples are then removed and the number of surviving target organisms can be counted and the rate at which they are killed ascertained

(Figure 44). *Capacity tests* are used on sterilants at the dilution at which they are being used and test by how much the sterilant can be polluted with micro-organisms and organic matter before its sterilant ability is lost and inoculated organisms survive. (See Test 3 on page 151.)

For details of all these methods see *Laboratory Methods in Food and Dairy Microbiology*, by W. F. Harrigan and M. E. McCance (Academic Press 1976).

# Chapter 10

# Principles and practice of cleaning

The absolute cleanliness of personnel, equipment and premises is very important at all stages in the handling of food – from the first stages of production of the raw ingredients, through transportation, processing, packaging and further transportation to retail and wholesale outlets, to cooking in homes, canteens and restaurants. Food must be protected from contamination by pathogenic organisms which can cause disease in consumers and from spoilage organisms which cause deterioration in the quality of the food. Additionally, good standards of cleanliness minimize the risk of rodent and insect infestation. It was for these reasons that food hygiene regulations were formulated (an outline of the main points of current legislation is contained in Appendix 1). The Food Hygiene (General) Regulations, 1970 do not, however, describe how desirable levels of cleanliness are to be achieved and recognized, although guidelines are laid out in various codes of practice (see Appendix 1, page 184).

As well as following guidelines, food processors adopt specific methods to monitor and control the quality of the foods they handle and sell. The types of method used are discussed in Chapter 9.

The pathogenic and food spoilage organisms may originate from the food itself, or from water used in processing, or from dust, dirt and debris associated with the food, or from handling processes or the handlers or from rodents or insects. These contaminating organisms include a very wide variety, individual foods tending to contain certain species. Waste food materials (known as 'food soil' when adhering to equipment and plant) harbour these organisms. As a result, used plant and equipment become contaminated, and, if not properly cleaned before further processing operations, act as a source of organisms, cross-contaminating other foods.

## Objectives of cleaning procedures

The first important objective of all cleaning procedures is to minimize the risk of cross-contamination. This applies as much in food processing areas as it does in canteens and food shops.

The second objective which is of almost equal importance in food processing is to ensure that deposits in the form of food soil and scale (derived from cleaning water and baked-on foods) do not build up. These deposits as well as harbouring micro-organisms also prevent the efficient functioning of equipment by impairing heat transfer mechanisms and can reduce the working life of the equipment through corrosion. This may happen, for example, with heat exchanger plates where the deposition of milk stone (see page 147) has been permitted.

## Definitions

Certain terms are used frequently in the context of cleaning procedures. The definitions used below are widely though not yet universally accepted.

*Detergent*. A substance, or mixture of substances, which when added to water helps to remove dirt and grease. Any detergent used in the food industry must be easy to wash away and must be non-toxic.

*Detergent/sanitizer*. Some detergents are combined with sanitizers to form one mixture which, when used, enables a reduction in the total number of stages in cleaning. The disadavantage of such combinations is that some of the sanitizing action is lost because it is inactivated by organic soil in the water. This does not occur when detergent is used, is rinsed away, and a sanitizer is used.

*Disinfectant*. See *sanitizer*.

*Disinfection*. The chemical destruction of micro-organisms, for example by phenolic compounds, but not usually bacterial spores. Disinfection does not necessarily kill all micro-organisms but reduces them to acceptable* levels.

The term 'disinfection' is used in non-medical contexts to describe the treatment of inanimate objects, but is used very little today in relation to food contact surfaces. However, it is used in relation to the treatment of the skin and body membranes.

*Food soil*. In food processing, food soil may be sugar(s), fats, carbohydrates such as starches, proteins or mineral salts, occurring in pure, mixed or very mixed deposits. The deposits change in character, tending to burn and caramelize, or dry on the surfaces of the equipment and this increases the difficulty of removing them. In catering, food soil also includes raw foods and fragments of cooked foods of mixed composition too, all of which tend to adhere more strongly on heating. Examples of food soils and the ease with which they can be removed from surfaces are shown in Table 31 on page 149.

*Milk stone*. A problem in the dairy industry. It is a deposit derived from the thermal processing of milk.

*Sanitizer*. Any chemical used in sanitizing or disinfection.

*Sanitizing*. Processes, chemical and other, applied both with and without heat, which cleanse plant, etc., resulting in microbial counts that are at acceptable*, low levels. It is a more comprehensive term than disinfection.

*Scale*. This is the deposition of mineral salts from hard water. Scale can harbour micro-organisms and permit underlying corrosion on stainless steel and impair heat transfer processes.

*Sterilant (sterilizer)*. An agent used in sterilization which destroys microbial life, including bacterial spores, and other living bodies. In practice, chemicals rarely fulfil the role, and in food processing high pressure steam and pressurized hot water are the only examples of sterilants.

*Sterilization*. The process of destroying or removing all microbial (and other) life. This is rarely the objective in cleaning routines in food processing and handling. Only where sterile plant is required – as in the treatment of ultra high temperature (UHT) milk and other foods – is sterilization used.

## Cleaning principles

Cleaning processes aim to achieve clean, hygienic surfaces.

### What is a clean surface?

A surface is clean when it is free from residual film or soil capable of contaminating food products which come into contact with it. Assessment of cleanliness, when carried out on the spot, involves three aspects:

1  *Sight* – under good light there should be no sign of soil on a wet or dry surface, and on a smooth surface draining water should not 'break' excessively.

---

* Acceptability criteria will be determined by quality control staff who should monitor sanitizing processes. After sanitizing, the surface treated should not significantly contaminate the food which subsequently comes into contact with it.

2  *Touch* – there should be no greasy or rough feeling to clean fingers rubbed over the surface. A clean white tissue wiped several times over the surface should show no discolouration.

3  *Smell* – there should be no objectionable odour present.

The results of the tests carried out should show that the levels and types of bacteria and other micro-organisms present would not cause serious recontamination of food which comes into contact with that surface.

It should be noted that *surface cleanliness* is not the same as *surface sterility* which, in most food processing and catering operations, would be impractical, expensive, unnecessary and impossible to achieve.

### Microbiological tests for surface cleanliness

#### Swabbing

A swab comprises a piece of cotton wool twisted round one end of a small stick, and it must be sterile before use. Immediately before sampling, it is moistened in sterile fluid and the excess fluid is removed by pressing the swab against the side of the container. The damp swab is then systematically and firmly rubbed over the test area of at least $100 \text{ cm}^2$ so as to pick up the organisms adhering to the surface. The swab is then replaced in a measured volume of sterile fluid (10 ml) and agitated to remove the organisms from the swab. (The end of the stick handled by the operator is broken off just before the swab is dropped into the suspending fluid.) Samples of the suspending fluid are then inoculated into melted, cooled, nutrient agar medium and incubated, after which, the solidified medium is examined for colonies which are of significance and these are counted. The significant organisms might be coliforms, pseudomonads, or a total viable count of all types present might be required. If the test is standardized (that is, done in the same way each time) it can indicate if cleaning techniques are successful or are failing to keep the numbers of contaminating organisms down to an acceptable level.

For example, a dairy surface may be routinely cleaned and regular swab tests may normally show that the microbial count after cleaning is less than $10^4$ colonies per $1000 \text{ cm}^2$ (an acceptable level). If one day the test shows that the count is, say, $3 \times 10^4$ per $1000 \text{ cm}^2$, this is no longer acceptable and it indicates that the cleaning has failed in some way – perhaps due to a new operator on the plant, or a new cleaning procedure being used, or the wrong concentration of sterilant. Whatever the reason, it must be found and rectified – further swab tests will indicate when the fault in the procedure has been corrected.

#### Rinse techniques

The organisms attached to a surface can be removed by rinsing it with a sterile rinsing solution. This may involve filling the object with the rinsing solution, as would be necessary when the insides of bottles and similar articles are tested, or running the rinsing solution over a surface, or through a section of pipeline or plant, as would be necessary in the assessment of CIP treatments. In either case subsamples of the rinsing fluid would be analysed for their total viable count of organisms. The results would be recorded as a count per object (bottles, buckets, etc.) or per area of plant surface.

#### Contact plates

A contact plate has a slightly raised surface of sterile solidified agar medium with a known surface area which can be directly impressed on to the surface of a test item. Organisms adhere to it and when it is appropriately incubated the organisms develop into colonies which can be counted. The disadvantage of such plates is that they cannot be used to investigate irregularly shaped surfaces and corners.

None of the three techniques described above remove all the organisms from the surface. There is no guarantee that the same proportion of organisms will be picked up each time a test is done. Also, in swab tests, organisms picked up from the surface may remain on the swab. Strict standardization of techniques does, however, permit comparison of routinely performed tests as standards can be set against model cleaning techniques.

## Cleaning routines

In choosing a cleaning routine, whether for food processing plant, for catering or retail equipment or for premises, four questions have to be asked and answered:

1 Will the cleaning process remove soil, including grease, fat, protein deposits and scale, whatever its condition?
2 Will the total process reduce the numbers of viable micro-organisms to defined 'acceptable' levels?
3 Are the cleaning detergents and sanitizing agents being used at the right concentrations and temperatures so that they are always as effective as expected? (See Tests 1, 2 and 3.)
4 Are the cleaning methods and chemicals suitable for the materials of which the plant and equipment are constructed?

Every cleaning routine has to follow the same broad lines of procedure to achieve a desirable level of cleanliness:
Removal of visible contamination
Removal of invisible contamination
Rinsing to remove chemicals

### Removal of visible contamination

All mechanical procedures, such as vacuuming, scrubbing, washing of surfaces, hot water spraying, scraping of organic debris from machinery, and other processes which physically remove and reduce visible dirt can actually reduce the total number of micro-organisms present. Otherwise the organisms present in the soil would, given enough time, multiply which could lead to a serious build up of contamination.

### Removal of invisible contamination

Sanitization, if successful, should leave only a few micro-organisms at 'acceptable' levels – a state defined by quality controllers and recognized by observation and tests. If the removal of visible soil is only partially successful, subsequent sanitization will probably also fail since the soil protects the organisms from the sanitizers and the soil may also reduce the effective properties of the sanitizing agent. Also, any heat applied at this stage will make the soil harder to remove. (See Table 31.)

### Rinsing

Rinsing is applied to remove all traces of sanitizer from any surfaces which will be in direct contact with food. The effects of the earlier cleaning procedures can be undone if the microbiological quality of the water used is poor. In this case, the water might act as a source of contamination, as can also happen with incorrectly constituted solutions of detergent or sanitizer.

Table 31   *Food soils and their removal*

| Component | Solubility on surface | Ease of removal | Change on heating |
|---|---|---|---|
| Sugar | Water soluble | Easy | Caramelization – more difficult to clean |
| Fat | Water insoluble, alkali soluble. | Difficult | Polymerization – more difficult to clean |
| Protein | Water insoluble, alkali soluble, slightly acid soluble. | Very difficult | Denaturation – much more difficult to clean |
| Mineral salts | Water solubility variable, but most are acid soluble | Easy to difficult | Generally easy to clean |

*Source: Design and uses of CIP systems in the dairy industry,* Document no. 117 (*International Dairy Federation (1979)*)

In broad terms a clean appearance is more likely to be safer for food handling than a dirty appearance, but it is not enough. Cleaning must extend to the areas which are difficult to see and clean, and all areas when tested must pass the microbiological and other standards set. Cleaning routines relating to plant and equipment are summarized in Table 32.

### Test 1   Use dilution test

| | |
|---|---|
| *Purpose* | To test which dilution of a sterilant should be used for a specific purpose. |

| | |
|---|---|
| *Method* | Laboratory cultures of target organisms* are exposed to a concentration of sterilant for a timed period which must relate to the conditions of normal use. After exposure, for a timed period (for example, 8 minutes) subsamples are taken and cultured in a nutrient medium to test whether any of the organisms have survived. (See also Figure 44.) |

| | |
|---|---|
| *Result* | *Growth* – use of the sterilant at this level will not ensure destruction of all organisms present.<br>*No growth* – the test organisms, at the cell concentration used, do not survive. It would be necessary to follow up this result with a capacity test (see Test 3). |

*Target organisms are those organisms which constitute a hazard and whose numbers are to be reduced by using the sterilant. Examples are *E. coli*, *Streptococcus lactis*, *Pseudomonads*, *Staphylococcus aureus* and possibly, in some circumstances, salmonellae. These, and other organisms, are the ones which would be used in testing sterilants for their effectiveness.

### Test 2   In use test

| | |
|---|---|
| *Purpose* | To check whether the solution of sterilant actually in use contains any, and if so how many, organisms. |

| | |
|---|---|
| *Reason* | When a sterilant is used at too low a dilution, the organisms which it is supposed to be eliminating survive and sometimes multiply. They may even be spread about when the sterilant is next used. This is an important factor when cleaning agents are recycled, as may occur in CIP processes. The result could be that levels of microbial contamination are increased rather than decreased. |

| | |
|---|---|
| *Method* | A sample of the sterilant is diluted to one-tenth by adding it to a solution which inactivates it. This prevents the sterilant from having any further inhibitory action. Then ten very small drops (each 0·02 ml) of this solution are dropped separately on to the surface of a well dried nutrient agar plate which is then incubated for up to 72 hours at 22°C. The plates are then examined for the presence of colonies derived from organisms present in the test sterilant solution. |

| | |
|---|---|
| *Result* | Any growth in more than five out of the ten drop tests indicates that the sterilant contains a significant number of viable organisms (the equivalent of 1000 recoverable organisms per ml). The sterilant is therefore ineffective and could well be acting as a source of organisms. |

### Test 3 Capacity test

*Purpose* To assess to what extent a sterilant can be soiled with micro-organisms and organic soil before it is inactivated.

*Method*  1 A suspension of target organisms (for example, *Pseudomonas fluorescens*, *E. coli*, *Klebsiella aerogenes*) at about $10^8$/ml is prepared.

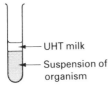

UHT milk

Suspension of organism

2 1 ml organic material (for example, UHT milk) is added to 10 ml organisms. The viable count is taken.

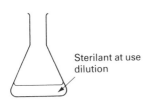

Sterilant at use dilution

3 6 ml sterilant at its use dilution is placed in a sterile flask.

4 1 ml of the organism suspension is added to (3) above, at time zero and at 10 minute intervals for 1 hour. On each addition the sterilant and the organisms are mixed by gentle swirling.

5 8 minutes after each addition of cells, 1 ml is removed. Five 0·02 ml drops are dropped on to the surface of a dried nutrient agar plate, and two drops are added separately to two nutrient broths containing an inactivator. Both plates and broths are incubated. Samples are taken after 8, 18, 28, 38, 48 and 58 minutes – by which time the sterilant is at approximately half its original strength.

*Result* After appropriate incubation (24 or 48 hours at 22°C, for example) the cultures are examined.

*The sterilant is still effective if fewer than five colonies from five drops are found, and less than two positive broths.*

*Note*: The concentration of sterilant, the number of bacteria added and the organic concentration can all be calculated. If at least three increments of organisms can be added before positive results are obtained the use dilution is considered adequate.

Table 32  *Cleaning routines in plant and equipment*

| Operation | Function |
|---|---|
| 1  *Thorough cleansing-removal of visible soil and dirt* | |
| Pre-rinse | Rinse with water to remove loose soil |
| Cleaning | Removal of all residual soil by a suitable detergent |
| Inter-rinse | The removal and rinsing away of all detergent and soil |
| 2  *Removal of invisible contamination* | |
| Sanitization | The destruction of most micro-organisms by the application of chemicals with and without heat |
| 3  *Post-rinse* | |
| Final rinse | Removal of sanitizing chemicals from the system with water of suitable micro-biological quality |

## Water: its qualities and uses in cleaning

Since water is used in most cleaning operations and as a final rinse, its quality affects final cleanliness. Suffice it to say here that cleaning water must be of good chemical and microbiological quality. The degree of water hardness affects the choice and cost of detergents which are usually more powerful in soft water. (Water is discussed in more detail in Chapter 11.)

Steam and hot water are among the best methods of sanitizing, when conditions permit, but can prove more expensive than chemical disinfection. Also, plastics and large items are not easily treated. The level of microbial destruction depends on the temperature reached, the humidity and the time for which the temperature is held. The efficiency of the process depends on adequate penetration of heat to all parts of the equipment.

### Simple boiling

This can be used to sanitize small items which will not be damaged by the temperature – such as cloths, jelly bags, knives, other implements and small utensils. Exposure of the item to boiling water (100°C) for 10 minutes destroys all vegetative cells and some of the spores present – only the more heat resistant spores survive. Where as immersion may be satisfactory in some cases, pouring so called 'boiling water' into buckets and similar utensils is not a reliable method of sterilization because of the difficulty of ensuring that the utensils have been adequately heated.

### Hot water

In a food factory, hot water at 60 to 65°C, sometimes containing suitable chemicals and applied with high pressure jets, is a very useful means of cleaning non-absorbant, uncomplicated surfaces such as tiles, floors and walls. The pressure of the water has the effect of removing organic soil attached to the surface and therefore reducing the microbial numbers by a (primarily) mechanical effect. Hot water can also be pumped through some processing equipment where it has the same effect. The heat of the water may also destroy organisms.

The common time/temperature combination in cleaning plant used for milk, milk products and ice-cream is 85°C for 10 minutes. Obviously a higher temperature is necessary if a shorter time is used, but this is often impractical. Hot water treatment destroys only the most heat sensitive of bacterial spores. In catering operations, water at a temperature of not less than 77°C is used to rinse food utensils after cleaning and washing up. The hot water then evaporates and this obviates the need to dry items with cloths.

### Steam

Steam can be used to sanitize food manufacturing and handling plant on site. It possesses the advantage that it has good penetrating power and can reach into corners in equipment to which access would otherwise be difficult. Generally steam is not used alone, but as a final sanitizing step after plant has been washed and cleaned with a detergent. The recommended time of application depends on the methods used. For example, for milk cans, steaming on a steaming stool for 1–2 minutes, depending on the capacity of the can, is recommended as adequate. For milking equipment or milk bottles steaming in a steam chest should be continued for at least 10 minutes after the temperature in the chest reaches 96°C. For storage tanks and other large vessels, steam should be injected until the condensed water issuing from the outlet of the tanks reaches 85°C, and then should be continued for at least 10 minutes. Steam at atmospheric pressure will not destroy all bacterial spores, but is effective against vegetative bacteria, yeasts and moulds, if applied as recommended.

The disadvantages of steam are that it is expensive because it requires special steam generating plant, it can be dangerous to personnel and it can cause damage to rubber seals and hoses. Escaping steam, for example, causes a steamy atmosphere which can damage paintwork, electrical fittings, etc.

### Autoclaving, or steam under pressure

Steam generated under a pressure which is greater than atmospheric pressure at sea level has a temperature higher than 100°C. At 121°C for an exposure time of 15 minutes there are very few types of heat resistant micro-organisms which can survive. Frequently times of much less than 15 minutes will very adequately destroy the contained micro-flora.

### Cold water flushing

After cleaning operations, detergents and sanitizers are flushed away using cold water. Since this is the last step in cleaning, good microbiological quality water is essential, otherwise the water will reinfect the plant.

## Detergents

These have been defined briefly earlier in this chapter. Detergents are available as either solids or liquids, the solid forms tending to be stable over longer time periods.

Some combinations of detergents and sanitizers provide balanced products which clean and sanitize simultaneously in a single cleaning process. Such products are used primarily for cleaning operations in catering, or for cleaning in place (CIP) where soil loads are light. Care must be taken to avoid the rapid inactivation of the sanitizing action of the product by organic soil in the cleaning water. The desirable characteristics of an effective cleaning agent are outlined in Table 33. Unfortunately, no single substance possesses all these characteristics, so commercially constituted detergents are mixtures of chemicals which, as a group, possess all the requirements. Table 34 indicates the more commonly used chemicals which have some detergent properties and are used in formulations for food processing plant and catering uses.

## Sanitizers

The efficiency of all sanitizing agents is influenced by concentration, contact time, organic matter and soil, pH, water hardness, combinations with detergents and the types of organisms present. All chemical sanitizers lack penetrative power, and therefore micro-organisms buried in soils may survive prolonged treatment. As

Table 33 *The main characteristics of an effective cleaning solution*

| Characteristic | Function |
| --- | --- |
| Organic dissolving power | The ability to make proteins and fats soluble |
| Wetting power | This property reduces surface tension of solutions and aids penetration into soils |
| Dispersing and suspending power | The ability to bring into, and keep in suspension, undissolved soiling matter, preventing redeposition on the clean surface |
| Rinsing power | This property ensures that deposits and solutions can be rinsed away |
| Sequestering power | Chemicals which combine with calcium and magnesium salts to form water soluble compounds, and thereby aid detergent action, reducing the risk of scale formation through precipitation |
| Bactericidal power | The power of an agent to kill bacteria (not necessarily spores) under defined conditions. This provides additional protection if cleaning is not 100 per cent effective |
| Buffering action | The ability to stabilize pH value under cleaning conditions |

normally used, chemical sanitizing agents are not effective against bacterial spores and should not be relied upon to kill mould spores.

As mentioned above, appropriate mixing of detergents and sanitizers may provide balanced products which clean and sanitize in one action. Such products may be preferred for applications where it is possible to achieve effective cleaning at moderate temperatures – for example, at temperatures below 60°C and where the soil load is light. At 70°C and above, detergent solutions used alone will kill most spoilage and pathogenic bacteria, and the use of a chemical sterilant may be unnecessary.

Table 34   *Detergent raw materials*

| Material | Examples | Uses |
|---|---|---|
| 1  Alkali | Inorganic alkalis<br>  sodium hydroxide<br>    (caustic soda) | High alkalinity, good buffering and rinsing power |
|  |   sodium orthosilicate<br>  sodium metasilicate |  |
|  |   sodium carbonate<br>    (soda ash, soda crystals)<br>  sodium hydrogen carbonate<br>    (sodium bicarbonate) | Low alkalinity – used in detergents which come into contact with the skin |
| 2  Phosphates | Orthophosphates<br>  trisodium phosphate (TSP) | High alkalinity, good buffering and rinsing power |
|  |   monosodium dihydrogen phosphate<br>  disodium hydrogen phosphate | Moderate alkalinity |
|  | Polyphosphates<br>  sodium pyrophosphate<br>  tripolyphosphate<br>  polymetaphosphate | Good water softeners, peptizing agents, deflocculants, dispersing power, suspending qualities and emulsifying agents |
| 3  Acids | Inorganic<br>  nitric acid<br>  phosphoric acid<br>  sulphamic aicd | Used to dissolve carbonate scale and some mineral deposits, often the effervescence with carbonate removes co-precipitated protein and fat |
|  | Organic<br>  hydroxy-acetic acid<br>  gluconic acid | Acids used in formulations to remove tenacious soil such as milk stone |
| 4  Sequestering agents | EDTA and its salts<br>Gluconic acid and its salts | Prevent precipitation of water hardness by chelating calcium and magnesium; they enable the formulations of detergent/sterilants to be heat stable and compatible with quaternary ammonium compounds (QAC) |
|  | Polyphosphates | Prevent precipitation of water hardness. They are not heat stable and not stable long term in alkaline solution and slowly lose their chelating power |

| Material | Examples | Uses |
|---|---|---|
| 5 Suspending agents | Starch<br>Sodium carboxymethyl cellulose | Assist in keeping undissolved soil suspended |
| 6 Inhibitors | | Minimize the corrosive attack of acids and alkalis on metals |
| | Sodium sulphite | Protect tin surfaces from attack by alkalis |
| | Sodium silicate | Protect aluminium and its alloys from attack by milk alkalis |
| 7 Anti-foaming agents | | Added to minimize foaming |
| 8 Surfactants | Anionic wetting agents<br>  sulphated alcohols<br>  alkyl aryl sulphonates<br><br>Non-ionic wetting agents<br>  ethylene oxide | Reduce the surface tension of the aqueous medium and promote good liquid/soil surface interfaces. Anionic and cationic agents are incompatible. Non-ionic wetting agents are compatible with both anionic and actionic agents |
| | Cationic agents – QAC | Often used as sanitizers |

Guidelines when using a sanitizer are:

1 *Never use* a sanitizer as a substitute for thorough cleaning.
2 *Only use* a sanitizer as an additional safeguard to thorough cleaning, and only when the sanitizer's action can achieve a positive benefit.
3 *Do not* use a sanitizer where sterilization is the object – substitute another method, such as the application of heat.
4 *Do not* use a sanitizer where more reliable methods are available – for example linen and crockery may readily be treated by heat at relatively low temperatures as part of routine laundering or washing up.
5 *Do not* use sanitizers when they are unnecessary. Experts consider it doubtful that they are needed for most domestic purposes, for example.

Sanitizers have to be selected from the very wide range available on the market. Without technical knowledge of the action of the constituents and without knowledge of the exact types of target organisms, it is well to be guided by the information given by reputable manufacturers. Where laboratory facilities are available, tests can show which of the available sanitizers is most suitable for the task in hand (see Chapter 9, Tests 1, 2 and 3). A sanitizer used in the food industry must have the following qualities:

1 It must have good bactericidal activity, be active against a wide variety of organisms and suit the product and the process for which the equipment is used.
2 It must have low toxicity to human beings.
3 It must not affect the colour, odour or flavour of food processed in the equipment being sanitized (for these reasons disinfec-

tants especially phenolic disinfectants, are shunned by the food industry – combination of phenolic compounds with chlorine produces chlorophenols which have a remarkable ability to taint foods, even at concentrations as low as one or two parts per billion; such tainting can lead to the waste of very large quantities of foods).

4   It must be cheap, effective and convenient.
5   It must be easy to dispense and wash away, and must be suitable for the method of application (manual, spray, cleaning in place).
6   It must not adversely affect the operators who handle it.
7   It must not damage equipment treated with it – that is, it must be suitable to the type of material of which the equipment is constructed.

These qualities consitute the ideal. Obviously, not all chemicals possess all these desirable qualities. Some of the more commonly used sanitizers are listed below – each of which possesses some of the desirable qualities.

## Chlorine

Sodium hypochlorite and other chemicals such as chlorinated trisodium phosphate tend to be used alone as sanitizers. They can be added by the user to suitable detergents to make detergent/sanitizers. Organic chlorine release agents, such as di-chloro-dimethyl hydantoin and sodium dichloro-iso-cyanurate, are generally mixed by the manufacturer with detergents to make detergent/sanitizers.

Available chlorine released from the above chemicals reacts rapidly with, and is inactivated by, residual organic matter. On storage, used chlorine solutions show a rapid decline in the available chlorine and so only freshly prepared solutions should be used. For most purposes, a cold solution containing 100 mg/litre of available chlorine, with an exposure of 15 minutes, is adequate. At temperatures greater than 40°C, and with chlorine concentrations greater than 200 mg/litre, damage to plant can occur because chlorine is corrosive to most metals, particularly

under such conditions. Chlorine is least corrosive in alkaline solution, at low temperature and with short contact time.

## Quaternary ammonium compounds (QAC)

These are supplied by manufacturers mixed with compatible detergents to provide detergent/sanitizers. Their bactericidal action is enhanced in alkaline conditions. They are only mildly corrosive to metals and can be used at higher temperatures and for longer periods than hypochlorite, to clean away any resistant food soil. In normal use conditions – 150–250 mg per litre QAC, at temperatures over 40°C, and a contact time of more than 2 minutes – most bacteria are destroyed. At lower temperatures or in more dilute solutions, some gram negative bacteria can survive.

## Iodophors

Iodine may be dissolved in certain organic and acidic chemicals to form products known as iodophors. In this form iodine is bactericidal in *acid* conditions. The acidity is usually provided by phosphoric acid. The recommended concentration of 50–70 mg per litre of free iodine yields pH values of less than 3 in water of moderate alkaline hardness. Iodophors stain plastics, and organic and mineral soil yellow.

## Sodium hydroxide (caustic soda)

This has good bactericidal properties. A solution of 1.5–2.0 per cent w/v at 45°C is effective in 2 minutes against most sporing bacteria. Sodium hydroxide is included in most bottle washing detergents and, when used at the concentration and temperature recommended, is highly bactericidal.

## Formaldehyde

This is used to fumigate rooms, for example it is used as a fungicide in cheese rooms.

## Cleaning practice

In practice, cleaning procedures can only operate with success if careful thought is applied to the design of the premises and the equipment installed and to the materials of which these are constructed. Also, the risk of cross-contamination will be considerably increased if waste and effluent are not disposed of effectively. These points are considered below.

### Design and construction

#### Premises

Much contamination of food can be avoided if food is prepared in premises which have been well designed. This applies to the structure of the building, whether purpose-built or adapted, to siting and design of equipment and to the layout and nature of working surfaces.

Ideally, the construction and siting of food premises should take into account the following points in order to provide good working conditions for staff, to minimize the risk of infestation, to avoid the accumulation of waste food and soil and to facilitate cleaning.

1 Since rubbish dumps, sewers, manure heaps, lavatories, drains, etc. are likely to attract or be infested by insects and rodents, food premises ought not to be sited near any of these. However, except in the case of new premises, a management can exercise little control over such hazards (unless they come within the perimeter of the premises), and food premises affected in this way must rely on their own standards of operation for safety.

2 Internally the walls should be smooth, impervious, washable and of light colour so that any dirt is visible and easily washed off. Floors should be waterproof, non-slip, wear-resistant, free from cracks and depressions, and slope evenly towards drainage outlets. The drain covers in the building and outside should be in a good state of repair and should be well secured to prevent rodent entry. Appendix 2 which deals with rodent and insect infestations gives more details of how buildings can be proofed.

3 The premises should be carefully maintained in a good state of repair and there should be sufficient space for the processing and preparation of food. The materials of which

the building is constructed should be easy to clean and maintain and should be resistant to wear and corrosion. For these reasons wood and plasterboard are unacceptable as internal finishing materials.

4 Ceilings should be smooth and the rooms as free as possible from beams, ducts and service and power lines. The latter should be housed in rodent-proofed conduits designed to offer minimal harbourage to dirt and pests.

5 Good lighting in food processing and preparation areas and in wash-rooms and cloak-rooms not only makes work easier and safer for staff, but also ensures that dirt and food soil are easily visible and can therefore be removed. Poor lighting means that food soil can accumulate unnoticed and act as a reservoir of micro-organisms and also as a food source for insects and rodents.

6 Good ventilation ensures that cooking vapours and smells and steam are extracted so that the room does not become stuffy, steamy and humid, which make work less pleasant and more difficult. Six changes of air per hour should ensure well ventilated conditions. Poor ventilation permits the formation of condensation on cold surfaces and encourages mould growth on walls and ceilings. Air borne mould spores add to the microbial flora in the environment and increase the likelihood of food contamination and spoilage. Filters fitted to fresh air inlets prevent the entry of dust and insects.

7 The temperature of the processing and storage areas should be controlled to comply with the recommendations of the codes of practice regarding the food being handled. Water used in food premises should be safe for human consumption without the need for further treatment (see Chapter 11), and should be chemically suitable for the uses to which it is to be put.

#### Equipment

Food processing, food transportation and food preparation equipment all have to be reused many times. To make the use and cleaning of

equipment easy, the design should take into account the following points.
Equipment should:

Effectively do the task for which it is intended.
Possess adequate safeguards for operatives.
Be made of materials which resist corrosion and are durable.
Be designed hygienically, being as uncomplicated as possible with few crevices, dead ends, corners, pits and joins in which food soil can accumulate, and from which it is difficult to remove dirt.
Be of non-corrosive materials.
Be sited so that all parts are accessible for cleaning.

### Materials in contact with foods

Because of the constant problems of cross-contamination, all materials directly in contact with food during processing and in catering operations should be impervious and free of joins and crevices.

*Stainless steel* is widely used in food processing because of its many advantages. It is durable; it can be manufactured to give very smooth surfaces; sections of pipe and plate can be welded and the welds subsequently ground down to give crevice-free joints; it is a good engineering material; certain grades (food grades) are corrosion-resistant; in catering and butchery it provides non-absorbant preparation surfaces.

*Plastics*, as smooth laminates, can easily be cleaned and are good materials for table top surfaces and for wall surfaces. In machinery some types of plastic are used for flanges and spray nozzles, and plastic is also used in conveyor belts, in food handling and processing equipment. Because of low surface hardness plastics tend to become damaged by continuous abrasion or by exposure to detergents and heat and their hygienic life is limited by this factor.

*Glass* – heat resistant glass such as Pyrex is sometimes used in pipe-work in food processing plant. It has the advantage – besides presenting a smooth surface – of being transparent permitting observation of the contained product. Small sections of glass tubing are often installed in stainless steel equipment as sight glasses for visually checking liquid levels, and sight glass ports are installed in tanks for the same purpose. Glass tends not to be used more widely because of its vulnerability to breakage.

*Wood* – generally wood should not be allowed in contact with food at any stage in processing (except as initial packaging boxes, etc.). However, in butchery, wooden blocks still provide an excellent chopping surface. There are several modern durable substitutes now marketed which, while still viewed by the butchery trade as slightly inferior chopping surfaces, do represent an advance in hygiene because they are less porous than wood and are easier to clean. Where wood is used it should be as join-free possible.

### Cleaning and sanitizing procedures – catering

### Crockery and cutlery

*The two sink system.* First, all food particles should be removed by scraping or washing the utensil under a tap. The food scraps should not be allowed to clog the drainage system, but should be collected in a strainer. The first sink is the washing sink; the water should contain a detergent and be at approximately 46–50°C. The detergent should be added to the wash water in the quantity recommended by the manufacturer and its action aided by the mechanical removal of scraps. For this purpose a mop or cloth can be used provided it is sterilized by boiling after use. The utensils are then transferred to the second sink which contains water maintained at a temperature of at least 77°C which is too hot for the hands. A loading basket should be provided in between the two sinks in which to stack the crockery, so that it can be placed in the sterilizing sink. The crockery should undergo a 'sterilizing' rinse for at least 1–2 minutes. (The term 'sterilizing' is used loosely here – it does not mean freeing the crockery of all microorganisms, but freeing it of pathogenic and other heat sensitive organisms.) The heat which the utensils absorb during this time is sufficient to dry them when they are taken out of the water and so there is no need to dry them with a towel which exposes them to the risk of recontamination. All crockery and cutlery can be passed through the two sink system and if there is any need to polish cutlery to get rid of water marks, this can be done with a clean dry cloth or disposable paper towel.

All electrically operated machines, such as can openers and slicing machines, should be switched off at the mains and the plug pulled out before the detachable parts are removed for cleaning.

*The three sink system* uses an additional sink for pre-rinsing before the washing sink is used.

Most catering premises have dishwashing machines for crockery and cutlery and, provided the manufacturer's instructions are followed, dishwashers are as effective as the two or three sink systems. Once crockery, cutlery and utensils are cleaned, they should be stored where they will be free from dust. Any equipment which is chipped or cracked should be replaced as the crevices will permit food soil to lodge and harbour micro-organisms.

*Kitchen and storage areas*
The main points to observe are shown in Table 35.

Table 35 *Cleaning of premises: kitchen and storage areas*

| *Item* | *Mode of cleaning* | *Frequency* |
| --- | --- | --- |
| Floors | Wash or scrub floor with hot water containing detergent, rinse with clean water | Wash at least once a day, supplemented by sweeping and wiping of spilt material during the day |
| Walls | Wash walls with hot water containing detergent. Wipe and rinse off | Weekly |
| Smooth working surfaces such as stainless steel, plastic laminate, marble | Remove all scraps of food then wash in hot water containing a detergent. Rinse with clean water. This can be followed by wiping with a disinfectant and a final rinse | After every operation |
| Absorbent working surfaces, for example meat blocks, chopping boards, and other wooden surfaces (see Food Hygiene Code of Practice no. 8) | Remove all scraps of food by scrubbing. Wash in hot water containing a detergent, rinse with clean water, wipe dry | Immediately after use |
| Equipment – mixers and slicers, etc. | Remove all detachable parts, and all food residues. Wash using the two sink system. Non-detachable parts should have all food particles removed and be washed in hot water containing a detergent. Rinse in water containing a disinfectant. Rinse in water | Between each operation |
| Dishcloths, wiping cloths, cleaning cloths | Boil in water, possibly containing detergent. Rinse and dry | Daily |

(a)    various designs of spray balls

(b)    spray ball operating at low pressure;
       high pressure would be applied in use

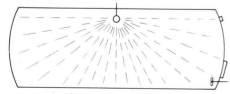

(c)    an example of a spray ball used in CIP of
       a tank

Figure 71   *Spray balls in CIP*

Table 36   *Waste receptacles and toilets*

| Item | Method of cleaning | Frequency |
| --- | --- | --- |
| Bins from the kitchen | Wash or scrub with hot water containing a detergent; rinse in water, possibly containing a disinfectant | Daily |
| Bins from the yard | Scrub with hot water containing a detergent; rinse with water, possibly containing a disinfectant | Once a week, or more frequently if possible |
| Concrete yard | Hose with cold water | Daily |
| Drains and gullies | Hose with cold water | Daily |
| Wash hand basins | Sprinkle with a foaming scouring agent containing bleaching agent which kills bacteria; wipe with a cloth | Daily |
| Toilets | Flush the toilet and sprinkle with one of the powdered cleansing agents; leave for a few hours if possible, before flushing | Daily |

**Waste receptacles and toilets**
See Table 36.

### Cleaning and sanitizing procedures – food processing

Cleaning in place (CIP) by the circulation of detergents and chemical sterilants has largely replaced hand cleaning in dairies, breweries and other types of food processing plant. It has many advantages including improved hygiene because the cleaning schedules can be adhered to more precisely and are less prone to human error.

The method consists of circulating water and various chemical solutions through process equipment. The combined effects of the turbu-

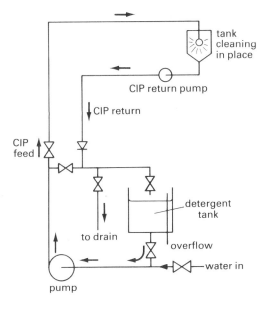

(a) a simple CIP unit

Figure 72  *CIP units and pipes*

(b)  CIP unit showing pumps, heat exchangers and detergent tanks

lance of the solutions and the chemical action remove the soil and the micro-organisms from the surfaces of the equipment. Satisfactory results depend on the temperature, time of exposure, the concentrations of the chemicals used and the physical action of the circulating fluids. Obviously it is essential that the cleaning solutions reach every part of the plant which has been in contact with the food product.

In closed equipment, such as pipelines and heat exchangers, the solutions have to circulate with sufficient turbulence and velocity to remove the soil load completely. All sections of the pipeline must be filled with solution, without leaving any trapped air and with the circulating fluids moving at 1–2 metres per second. It is not necessary to fill large vessels (such as milk silos or fermenting vessels) for cleaning in place. All that is required is to ensure that all internal surfaces are covered by cleaning solutions. Static spray heads or rotating jet devices fulfil this requirement (see Figure 71).

Special additional plant has to be added to the food processing plant – detergent storage tanks,

(c)  view from above of part of a dairy UHT-processing plant showing the number of pipes which have to be cleaned by CIP chemical metering pumps and circulating pumps. The complexity of CIP systems depends on the requirements of the food processor and can range from very simple systems to highly complex, totally computer-controlled systems. Figure 72 shows a plan of a simple installation which incorporates all the elements (in principle) of any system.

Some CIP systems depend on the single use of the detergent for one circulation clean.

Other systems, because all the detergent power of the detergent formulation is not lost after one clean, are designed to recover the detergent and reuse it in subsequent cleaning cycles. In reuse systems, monitoring devices in the system check whether the chemical reagents are circulating at the correct strength, then feed back that information to the metering pumps, which will, if necessary, allow additional chemical into the system. Reuse systems also permit water recovery, whereas in single use systems water and detergent go to waste after use.

For more information about CIP systems, a useful reference is T. C. Tamplin's 'Cleaning in place (CIP) systems and associated technology' in *Developments in soft drinks technology*, Volume 2 (Applied Science Publishers Ltd 1981).

# Chapter 11

# Water in food operations

This chapter is concerned with the microbiological aspects of water, how the spread of disease through the use of infected water is avoided, and how water quality relates to product quality in food processing.

When water, such as a river or the sea, is polluted with sewage it becomes contaminated with human faecal pathogens and other organisms. These gradually die out and if the water is only intermittently polluted, it will be self cleansing, mainly as a result of the cells dying and dilution. But continuous pollution by sewage will obviously result in a continuous risk of contracting disease for those who use the water.

The provision of good quality drinking water to all people is one of the major objectives in world development today. To this end some countries are fortunate in having an abundant supply of water from deep wells and underground springs; others have to make extensive use of rivers, lakes and other sources. The supplies of drinking water derived from these sources must not constitute a danger to the health of consumers – either through infective organisms (bacteria, viruses, protozoans) or through the presence of toxic, carcinogenic or radioactive materials.

The continuous supply and distribution of good quality potable water is a major technical achievement, but one which has to be continuously and carefully monitored to check the maintainance of quality. Minimum international standards which can be achieved by all countries in the world have been agreed through the World Health Organisation (WHO), and from these minima it is hoped that world-wide standards for drinking water will be continuously raised. In some countries standards very much higher than the WHO international standards can already be readily attained.

Today, in the UK the vast majority of homes, offices, shops and business premises are connected to a mains water supply. This water is treated in such a way that it is safe for drinking and cooking and for all other domestic purposes. The mains water is thus supplied at potable quality that is, it is safe to use and drink. This means that its quality is equal to or better than the minimum requirement for safety, and hence is substantially free from pathogenic organisms and toxic chemicals but it does *not* mean that it is entirely free from all micro-organisms.

Fortunately it is rare for the mains supply to become infected and cause illness. But if, for any reason, sewage does contaminate the supply, as would probably occur in flooding, the risk of the water carrying diseases becomes high. In such circumstances the only possible temporary solution, other than not using the water, is for all water to be boiled or adequately chlorinated before it is used for any purpose.

Appreciation of how important water could be in the spread of disease began as long ago as the mid-1800s. For in 1854 John Snow recognized, when investigating an outbreak of cholera, that the drinking water used by those afflicated had a common source and was responsible for the outbreak in which many died. In 1856, William Budd concluded that typhoid fever could be spread by milk or water polluted by the excretions of infected persons. In the years that followed, microbiology, and particularly bacteriology, became established as a science and great progress was made in the understanding of diseases caused by bacteria. Then, in 1905, Alexander Houston initiated the chlorination of drinking water in the UK, which substantially helped to abolish water borne disease. And soon other countries adopted the practice, to their benefit. Today, as before, where drinking water purification is not applied, or fails through lack of control, outbreaks of disease spread by water occur. Two examples of breakdowns in control are given below.

In 1963 holiday makers returning from Zermatt contracted typhoid fever. Determined and thorough investigation showed that sewage containing the typhoid organisms had seeped into an undetected leak in the mains water pipe – from which source the tourists had obtained their water. This is an example of a direct link between pollution of water and subsequent disease. But the link may be more indirect. In a case in 1964, in Aberdeen, a number of people contracted typhoid fever from eating corned beef canned in Argentina. It was revealed through investigation that *Salmonella typhi*, the causative organism, had been able to enter through minute holes in faulty seams on some of the cans and had survived in the corned beef. The salmonellae had gained access because the cans of processed meat were cooled after processing in untreated river water (water of the river de la Plata) which was contaminated with both sewage and typhoid organisms. The cooling process created a lower pressure within the cans and polluted water was sucked in.

## Potable water

The principle observed in the supply of potable water is that the untreated supply should be of as high a quality as possible, and protected from all actual and potential sources of pollution to the greatest possible extent. In this country a lot of

drinking water is supplied from reservoirs in the mountainous regions of the country and from underground reservoirs and wells. But no matter how protected the source of water, a certain amount of treatment is required – possibly filtration, to remove suspended materials, and coagulation (by the addition of special coagulating chemicals), to help get rid of very fine particles and organic materials in solution. The coagulated material is then separated out of the water. The water may also be treated to remove other substances such as oils, and agricultural fertilizers and pesticides, and will finally be chlorinated or otherwise treated to kill microorganisms which are still present in the water. The final chlorination level of the water, as discharged, is quite low, and not high enough to destroy any invading organisms seeping into the pipe work. The quality of the water has to be protected by the safety of the distribution system. The aim is to supply water which is free from pathogenic organisms and from concentrations of chemical substances that may endanger human health. It is also the aim to make the water as pleasant as possible – in the sense of coolness, absence of turbidity, absence of colour and of any disagreeable taste or smell.

The amount of treatment that water receives varies with the level of need – so that daily or seasonal variations in raw quality do not affect the production of potable water of consistently high quality. But some raw supplies of drinking water are drawn from rivers into which waste material is discharged – sewage, industrial chemicals, oils, agricultural fertilizers, weedkillers, insecticides, etc. Where such water is used, obviously the treatment to raise it to potable quality is more extensive and problematic than when the raw supply is of good quality. Because major cities often have to use river water as a source for drinking water, it is not surprising that increasingly strict regional and national control over what substances may be discharged into the rivers is applied. In contrast, some domestic and other premises may use well water as a direct supply. Although not connected into the mains system such water must also consistently be of a quality equal to, or better than, the minimum standards required for potable water.

### Uses and abuses of potable water

Potable water in food premises where mains water is supplied can be trusted to be safe. It is used by employees for their personal use – for hand-washing, showers and for flushing of toilets – and it is used for washing foods, utensils, plates and cutlery; as an ingredient in many foods; for washing larger items of standing equipment; for washing walls and floors; and it is also used for washing overalls and cloths. The standard at which potable water must be supplied is given below, but it is important to note that the way the user uses water may undo all, or much, of the good the water authorities have done. For example, if the mains water runs into an uncovered header tank – often housed in a roof space – there is the possibility of birds, squirrels and other rodents, such as rats, drowning (and rotting) in the water. The water which emerges from the taps would then *not* be of the quality of the original supply. To give another example – water which comes out of the end of hose pipes may be inferior in quality compared with the mains supply simply because the hose pipes are left trailing on the floor and the nozzles are liable to pick up any mud, faeces, food particles, etc. with which they may come into contact. Even if the water is not contaminated, potable water may not be of a high enough standard to satisfy the needs of food processing.

### Minimum international standards for drinking water quality (WHO 1971)

#### Recommendation

Water circulating in the distribution system, whether treated or not, should not contain any organism that may be of faecal origin.

#### Standards

1   Piped supplies
    (a)   Water entering the distribution system
          (i) Chlorinated or otherwise disin-fected supplies

It should not be possible to demonstrate the presence of coliform organisms in any sample of 100 ml. If the test is positive for the organisms, an immediate investigation of the purification process and method of sampling should be carried out.

(ii) Non-disinfected supplies
There should be no *E. coli* in 100 ml. If *E. coli* is absent, up to 3 coliform organisms per 100 ml may be tolerated in occasional samples from established supplies, provided they have been regularly and frequently tested. If repeated samples show the presence of coliform organisms, steps should be taken to discover and, if possible, remove their source.

(b)   Water in the distribution system
Ideally all samples taken from the distribution system, including consumers' premises, should be free from coliform organisms. In practice this is not always attainable, and the following standards for water collected in the distribution system are therefore recommended:

(i) Throughout any year 95 per cent of samples should not contain any coliform organisms in 100 ml.

(ii) No sample should contain *E. coli* in 100 ml.

(iii) No sample should contain more than 10 coliform organisms per 100 ml.

(iv) Coliform organisms should not be detectable in 100 ml taken from two consecutive samples.

If any coliform organisms are found, the minimum action required is immediate resampling. The repeated finding of 1–10 coliform organisms in 100 ml, or the appearance of higher numbers in individual supplies, suggests that undesirable material is getting into the water and measures should be taken to discover and remove the source of the pollution.

2  Individual supplies

Although the standard above may not be attainable it should be aimed at and everything possible should be done to prevent pollution of the water. It should in any case be possible to reduce the coliform count to less than 10 per 100 ml.

## Food processing water

The incoming water supply to a food processing factory arrives at a certain microbiological and chemical standard and is destined for different categories of use, for which different minimum standards of water quality will be required.

Examples of different uses are:

1  Potable quality for the personal use of employees, and in the kitchens and canteens supplying food to the works.
2  For product manufacture – both in processing and in plant cleaning (CIP).

Besides supplying the needs of employees, much of the water in factories is used for hosing the walls and the outside surfaces of the plant and for cleaning vehicles, tankers, etc. Water may also be used in the form of sprays and jets to sluice floors, walls, grids, conveyor belts, etc. to clear them of organic soil (vegetable peelings, ice-cream over-run, meat pieces, etc.) or to flush away detergents or sterilants. The minimum standard that this water has to attain is the same as that required for potable water, which will be adequate for some purposes, but not when the water is coming into contact with the plant. In this case additional chemical and bacteriological standards will usually be set by the quality control department. For example, in dairies, the absence of or maximum acceptable limits for, species of *Pseudomonas*, *Achromobacter* and *Alcaligenes* would be specified since these are organisms which can grow at low temperature and spoil refrigerated dairy products.
3  For product cooling (plate heat exchangers, retorting, continuous sterilizing etc.).
4  For boiler feed water.

On site treatment of incoming water may therefore involve a suitable combination of settling, coagulation, filtration, ion exchange and activated carbon absorption – the complexity of treatment being determined both by the initial quality of the incoming water and the required standard for use. Suspended matter is undesirable because it blocks filters and introduces non-food materials. Certain dissolved minerals – ions such as $Fe^{2+}$, $Fe^{3+}$, $Mn^{2+}$ and $Ca^{2+}$ – can affect the food quality. For example, in vegetable processing a high $Ca^{2+}$ content makes the processing more difficult and produces vegetables with a hard, undesirable texture. Some salts can separate out of solution and deposit inside the plant producing scale – particularly in heat exchangers and in pipes. This is also a problem in boiler feed water. The build up of scale in boilers and heating pipes is an expensive problem – and so boiler feed water is usually softened before it is used. Scale in the plant adds to the problems of ensuring adequate cleaning and aids corrosion. Dissolved organic matter in the form of colours, tastes and odours can radically and adversely affect the quality of the product – for example, in the production of soft drinks where colour and flavour must be consistent. Dissolved gases are not always desirable – de-aeration of water is essential in the successful UHT processing of fruit juice, otherwise the product will deteriorate in storage. The micro-organisms in the incoming water can be very varied – including diatoms, fungi, bacteria, algae, rotifera, protozoa and nematodes. If these organisms do get into the food plant they and their decomposition products can give rise to 'off' odours and tastes, food borne illness, food poisoning, food spoilage or technical problems in the plant such as the build up of slimes or the blockage of sections of the plant.

In some cases micro-organisms, such as coliforms, Citrobacter, *Aeromonas hydrophila*, *Pseudomonas aeruginosa* and fungi, can grow in the water and in storage tanks, on nutrients provided by plumbing materials. It is known that leather which is used for washers supports bacterial growth, as can polyurethane jointing, polyvinylidene chloride (PVC) and many other materials which are based on vegetable oils, shellac, vegetable fibres or animal fats. However materials which are made from polytetra-fluoroethylene (PTFE) or silicone will probably not support bacterial growth. Today, use of the former group of materials is avoided in food plant; but in the past, before it was generally realized that such materials could support bacterial growth, problems were experienced both in the food and soft drinks industries causing the spoilage of products, in surgical units in hospitals, in medical equipment supplied to the home (such as kidney machines) and in public food vending machines. Thus, when samples are taken in order to assess the quality of supply, the samples must include ones taken from the point of use. Indeed when making assessments of water quality great care has to be taken both in setting microbiological standards and in choosing the sampling sites.

So water may be supplied at a potable quality but not at a quality good enough for plant rinsing. If this water were to be used as a final rinse after plant cleaning and sterilization it would 'unsterilize' the plant and thereby introduce organisms into the product. Where milk, for example, is inoculated with starter organisms to make curd cheese, unwanted organisms such as coliforms, if accidentally added from the plant, may grow alongside the starter organisms, produce gas and spoil the curd – with consequent financial losses. If the quality of the plant rinse water is not tested, because it is assumed to be all right, it may be some time before the problem with the product is solved.

An example was quoted earlier on page 164 of the danger of not using water of good quality for product cooling. In a similar way, cooling water (plus any organisms it may contain) can sometimes penetrate to products through minute leaks in the seals on plate heat exchangers. This should not happen, but sometimes does if plant is not maintained adequately.

It is estimated that, on average, 33 tonnes of water are used for every tonne of food produced. Even allowing for absorption into the product and evaporation, the bulk of the water

must eventually be discharged. The effluent water (waste) contains contaminants derived from its use. Today, it is not acceptable that untreated factory waste, especially that from food factories and agricultural units, should be discharged untreated, and usually a certain amount of cleansing of the effluent is demanded by the local authority. This is because the materials such water contains, both in suspension and solution, are very polluting to the environment.

Thus, food processing factories frequently find that the incoming water, even if of potable quality, may not be of high enough bacteriological or of suitable chemical quality for food processing; and that the untreated outgoing effluent is not of high enough quality to be acceptable to the local authority. They find, in fact, that they must treat both incoming and outgoing water.

## Microbiological assessment of water quality

The WHO publication *International Standards for Drinking Water* sets out very clearly the objectives, including numerical standards where appropriate, the test methods and the frequency of sampling required, regarding the bacteriological, virological, biological, radiological, physical and chemical aspects of drinking water. This is to ensure that water which is described as potable should have an assured level of safety from the possible hazards as outlined above. From the microbiological point of view, the standards relating to the detection of faecal pollution are the most important.

### Coliforms as indicators of water quality

The coliform group of organisms, which undoubtedly can and do occur in faeces, also survive well and grow in many environments outside the animal or human body. The usual practice is therefore to assume that if coliform organisms are found in water supplies they are of faecal origin, unless it can be proved otherwise – in other words, to err on the side of caution. The presence of coliforms, whatever the origin, does indicate an undesirable source of contamination.

### Detection of coliform organisms in water using the multiple tube technique and probability tables

#### Outline of the method
Tubes of MacConkey's broth are inoculated with a range of volumes of sample water and incubated at 37°C for up to 48 hours. The presumptive test looks for a change in colour in the growth medium from purple to yellow – showing production of acid (A) and gas (G) trapped in an inverted tube in the medium. Both the acid and the gas are due to the multiplication of the original organisms from the sample water, in the medium. The presumption is that coliform organisms present have brought about the changes. However, since anaerobic spore bearing organisms, unlike coliforms, can survive chlorination, positive tubes have to be checked to confirm the presence of coliforms and checked for the presence of E. coli (see Test 4).

From the combination of positive (A and G) and negative tubes the numbers of coliforms per 100 ml of the original water sampled can be estimated from probability tables (see Table 37).

#### E. coli as an indicator of water quality
Faeces contain hundreds of millions of coliform bacteria per gram – of which E. coli is an important member.

Bacterial pathogens may also be present but, by comparison, in relatively low numbers. Water infected with faeces becomes infected with E. coli organisms. Thus, when water is examined for possible faecal pollution it is actually examined for the presence of E. coli. If E. coli is absent from the 100 ml test volumes, then, by inference, pathogenic bacteria will also be absent from that volume and, because their numbers are so low in comparison to E. coli, they will also be absent from greater volumes – so the water can be considered safe from bacterial pathogens.

Test 4   *Detection of E. coli in water using the multiple tube technique*

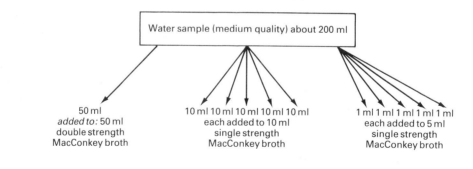

Water sample (medium quality) about 200 ml

| 50 ml *added to:* 50 ml double strength MacConkey broth | 10 ml 10 ml 10 ml 10 ml 10 ml each added to 10 ml single strength MacConkey broth | 1 ml 1 ml 1 ml 1 ml 1 ml each added to 5 ml single strength MacConkey broth |

*Principle*   Positive tube in the coliform detection test (see Table 37) should be tested for the presence of E.coli.

*Method*   Each positive tube is sub-cultured into two tubes of brilliant green bile broth*, and one is incubated at 37°C for up to 48 hours (to confirm the presence of coliforms), the other is incubated at 44°C and inspected after 6 and 24 hours for acid and gas

(E. coli but not coliforms can produce acid and gas at 44°C).

*Result*   Probability tables are used to estimate the numbers per 100 ml of E. coli in the original water. The presence of E. coli indicates recent faecal pollution.

*Lactose ricinoleate broth or MacConkey broth may also be used.

Table 37    *Multiple tube techniques: the range of volumes used depends on the estimated water quality:*

| | **Volumes of water likely to be tested** | | | | |
| | *50 ml* | *10 ml* | *1 ml* | *1 ml* *1/10* dilution | *1 ml* *1/100* dilution |
| --- | --- | --- | --- | --- | --- |
| *Estimated water quality* | *Number of replicate volumes* | | | | |
| Good | 1 | 5 | | | |
| Medium | 1 | 5 | 5 | | |
| Poor | 1* | 5 | 5 | 5 | 5* |

*Possibly included.

The presence of *E. coli* in water samples indicates recent faecal pollution because this organism does not survive long in the alien aqueous environment. Sometimes other indicator bacteria – faecal streptococci and spore forming organisms – are looked for in water samples. Because these organisms survive longer in the water than *E. coli* does, the results of such tests give a more complete picture of the history of the pollution of the water.

The methods by which the samples are taken and analysed have to be carefully standardized, so that over a period of time information about the regular quality and variations (daily and seasonal) is built up.

### Viruses

Faeces also contain pathogenic viruses. So where water is contaminated by sewage (and therefore faeces) it also contains pathogenic viruses. Test samples of water which are positive for *E. coli* are therefore liable to contain virus particles. But samples which are negative for *E. coli* could also contain virus particles because they can survive longer in water than *E. coli*. Viruses are not easily removed from water by conventional water treatment for plant and, to a certain extent, some resist chlorination. However, in practice it has been found that in water in which free chlorine is present if coliforms are absent viruses will be too. Exposure of the water to 0.5 mg per litre of free chlorine for 1 hour has been found in practice to be an effective treatment, although there is much room for further research into this area.

### Colony counts

Colony counts of the many organisms which can grow at 37 and 20°C can provide useful information on the quality of both raw and treated water. Data from samples taken at a series of points in water treatment plant help in the control of the process – there should be a progressive decrease in counts as the raw water is processed. The insides of the pipes of the water distribution system may be colonized with a biological slime layer which may contribute some bacteria to the water. Also, the packing in joints in pipes may be colonized and maintenance repair work may also add organisms. Samples from points along a distribution system should not show high microbial counts, but such samples may perhaps indicate a point of pollution upstream of the sample.

To count the organisms in the water may require their concentration. For this purpose a volume of the test sample can be filtered through a membrane with pores fine enough to hold back bacterial cells. The membrane can then be placed on a solid nutrient medium and the individual bacteria can then develop into colonies. Each colony which is formed represents a bacterial unit which was suspended in the original sample. The types of organisms detected in mains water supplies by this method (membrane filtration) do not usually affect the potable quality of the water. Sometimes, however, they cause an undesirable haze which should be attended to. The presence of any organisms detected, such as *Pseudomonads*, can be very undesirable and cause problems in food processing. (See Figure 73.)

In food processing water analysis, the technique of membrane filtration may be applicable for the detection of organisms relevant to the spoilage of the product or which cause problems in the running of the plant.

A useful reference for methods of analysis is the DHSS *The bacteriological examination of water supplies*, Reports on Public Health and Medical Subjects, no. 71 (HMSO 1969).

## Removal of organisms from processed water

Micro-organisms are not necessarily removed by the filtration and coagulation stages in the treatment of water and positive steps have to be taken to remove them. Heat is not normally used as a sterilizing agent because of the very considerable costs which would be incurred. Other physical and/or chemical methods are both effective and, relative to heat, inexpensive.

### Chlorination

Chlorination of water is a very effective method of reducing its microbial content – the sterilizing effect being brought about by the presence of the free chloride ions. For this purpose, liquid chlorine is used for large scale treatments, and either chlorine dioxide or hypochlorite for smaller scale treatments. The effectiveness of chlorination depends on the concentration used, the contact time, the pH, the temperature, the amount of organic matter present and the number of micro-organisms. The problem of this method is that the chlorine reacts very readily with organic matter dissolved in the water, and, as a result, the ability of the chlorine to sterilize the water is considerably reduced because the chlorine is 'bound' and so not readily available. Sufficient chlorine must therefore be added to the water to satisfy this reaction – 'the chlorine demand' – and to leave enough free residual chlorine over to effect rapid disinfection – a

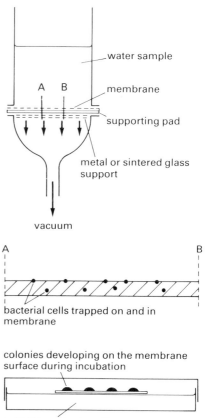

Figure 73  *The membrane filtration technique. Where organisms occur in low numbers per unit volume of fluid, and possibly would not be detected by other techniques, they can be concentrated by the filtration of large volumes. The cells are retained on the membrane and develop into visible colonies when the membrane is incubated on the surface of a suitable solid medium*

method called 'break point chlorination'. However, if water is contaminated with phenols and tars, chlorination produces compounds known as chlorophenols which give a very disagreeable taste to the water, even in minute quantities, and can cause tainting of food products.

An alternative method, called 'superchlorination', involves a much greater dose of chlorine being added than in the break-point chlorination method. Disinfection occurs and excess chlorine is removed from the water by contact with sulphur dioxide or by passage through activated carbon, leaving sterilized water.

When continuous chlorination is applied to water beyond the break point, the free chlorine level is usually set at between 5 and 7 mg per litre, and the water can then be used for the spray cleaning of intricate pieces of food handling and food manufacturing equipment, such as conveyors and conveyor belts, can coolers, product washers, flumes and fillers, peelers, dicers, etc. Chlorinated water is used either in spray form or for the immersion (where appropriate) of small items. It is *not* usual practice to use water containing chlorine at 5–7 mg per litre inside food plant because of the possibility of tainting the food product. A chlorine concentration of 5–7 mg per litre is very much higher than that used for mains water where the residual level is normally in the region of 0.2 mg per litre. For final rinsing purposes inside the plant, good quality water, with a chlorine level (if any) close to that of mains water, is used. The mains water level is insufficient to protect the water from contamination originating within the distribution system. Larger doses of chlorine, can give some protection in an emergency but the chlorine is then noticeable as a taint. Very high chlorination levels make water unacceptable for drinking and for final rinse purposes inside plant.

## Ozone

Interest in the use of ozone as a sterilant for water is increasing, particularly on the continent. It is considered to have superior bactericidal action to chlorine and creates no taste problems, in contrast to the risk of tainting by chlorophenols which may occur when chlorine is added to water. Also, once applied, ozone leaves behind no residual other than a high dissolved oxygen content.

## Bacterial filtration

Another way of removing organisms is to pass the organisms through filters fine enough to retain them. This is only effective when the levels of suspended material are very low indeed, otherwise the filter very rapidly becomes blocked.

## Ultra-violet light

UV light can be an effective microbicide. Since UV has very little penetrating power, it is most effective either for surface sterilization of objects or for treating clear liquids. It cannot be used with any certain effect for opaque or turbid liquids. The principle applied in water treatment is to allow UV light of microbicidal wavelength to penetrate thin layers of the water in which the micro-organisms are suspended.

Basically, a sterilizing unit comprises a lamp (low pressure mercury vapour) encased by an outer tube through which the water to be sterilized is pumped. The lamp emits a peak radiation close to 260 nm and the sterilizing unit is designed to ensure sufficient exposure time to each unit volume of water flowing through. This brings about a sterilizing effect. Used properly, these units can work well, but problems arise if the water to be treated is cloudy and if the suspended matter deposits on the lamps. The UV rays cannot readily penentrate the deposited layer and treat the water. In any case, regular cleaning of the sterilizing units is vital to ensure that the correct dosage of radiation reaches the water flowing through.

In food processing, UV sterilization of water can be very useful as an end treatment after water has been treated by chemical methods and the residual chemical has been removed. Since there is always the possibility of unwanted organisms intruding after treatment, the practice of UV sterilization immediately before water reaches use points acts as a safeguard to the water quality used in the food process and has the advantage of leaving no residual.

## Water associated microbiological problems in food processing

### Poultry

In the mass processing of poultry very large numbers of carcasses (7000–8000) can be dealt with per hour and water plays an important part in preparing the meat (see Figure 74). After the birds are stunned and bled, the carcasses are scalded in water at a temperature of between 53 and 61°C. This is followed by dry or wet de-feathering in a plucking machine. In wet de-feathering, up to a litre of water per carcass may be used. The plucked carcasses are then automatically eviscerated, cleaned (usually) and chilled. After chilling, which either involves immersion in cool flowing water for up to an hour or dry chilling in air, the birds are packed and may be frozen.

Before processing, the skin of the birds is heavily contaminated with various organisms, particularly Pseudomonas, Achromobacter, Acinetobacter, Flavobacterium, Corynebacteria, Aeromonas, Enterobacteriaceae (including salmonellae) lactic acid bacteria, Micrococcaceae and yeasts, which adhere to the skin of the chicken. In the processing of the birds, and especially in the wet processes, the load of organisms that each carcass carries is added to.

In order to use large volumes of water in the most economical manner, it is the practice to

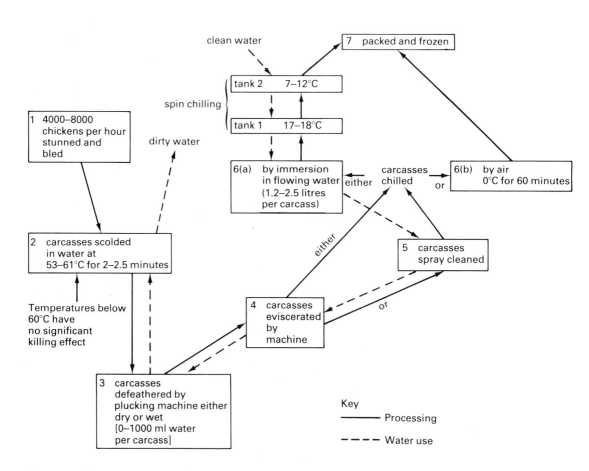

Figure 74  *Chicken processing*

apply the water in a counter direction to the progressive processing of the birds. The clean water is used for the last process in preparing the birds (see 6(a), Figure 74) and the used water is then fed on to the previous processing stage. Even if the water is used in this manner it inevitably picks up a bacterial load which contaminates the carcasses being processed and aids the spread of particular contaminants. In the hot water stages (see 2, Figure 74), temperatures below 60°C do little to reduce the microbial load the water carries. Also, even if the water is chlorinated, the high organic content inactivates much of the chlorine and runs the risk of not being very effective. Also, it has been shown that machinery, such as the pluckers, which remove the feathers, can harbour organisms in spite of the use of chlorinated water.

In view of the widely recognized problem of the infection of poultry carcasses with salmonellae (see Chapter 4) and the commercial need to produce carcasses which keep well at refrigeration temperatures, the fact that research has shown that when the use of water is cut down at every possible point in the process, the bacteriological quality of the birds produced is improved, is an incentive to the industry to change and improve its processing practice.

### Meat and abattoirs

In a similar way to poultry, meat carcasses carry a bacterial load. It has long been established that the hides and the water used to wash the carcasses are sources of both mesophilic and psychrophilic bacteria, particularly Achromobacter, Flavobacter and Pseudomonas species – all of which are capable of growing at low temperatures. The intestines of the animals can provide a source of organisms if the carcasses are not gutted properly, particularly, of course, Enterobacteriaceae (including salmonellae). The wet floors and walls of chillers provide an environment which favours the growth of gram negative psychrophilic spoilage organisms (Pseudomonas, Acinetobacter, Moraxella and Enterobacteriaceae). It would, however, be difficult to eliminate these organisms from these

sites and, provided good handling practices do not allow cross contamination from the floors, etc. to the carcasses, elimination of such organisms should not be necessary. Conversely, if such organisms are permitted in some way to contaminate the water which is used to wash the carcasses, then the microbial load of the meat would be liable to increase.

### Canning

In canning, cooling water is used in heat exchangers and for direct spraying or immersion of the cans, following retorting. In these cases the removal of colours, tastes and odours from the water is not important; but the water should be de-ionised in order to reduce the problems associated with scale.

New container designs involving soldered tin plate, two piece cans, shallow or deep drawn aluminium containers and flexible retort pouches of three ply laminates (made of polyester, aluminium foil and heat stable polypropylene) are coming increasingly into use. However, spoilage of the product in the can or pouch can occur due to the entry of contaminants through pin point holes in the seams. Cooling causes a vacuum and therefore inward suction of the cooling water and any organisms it may contain. Examples of bacteria which are likely to cause problems are *Bacillus macerans* in the spoilage of canned fruits and *Bacillus polymyxa* in the spoilage of canned vegetables. Other organisms such as cocci and both sporing and non-sporing bacterial rods may also contaminate foods such as canned sauces and condensed milk. Yeasts and moulds only rarely represent a problem and then only when the containers are grossly defective. When gas-producing organisms contaminate cans they cause swelling, and the original leak may become sealed. Spoilage of canned foods can be due to several species at once or to one type. The latter may occur if chlorinated water has been used, which has permitted one species to survive while others have been destroyed.

Continuous chlorination by chlorine gas or by hypochlorite to give a residual of about 5 mg per litre can be effective. The odours and taints can be a problem if the high levels of chlorine demanded in break-point chlorination are used and the water contains a high organic content. It is important that the build up of the organic content, through microbial growth (for example, of Flavobacteria or algae) in the cooling water, is avoided by regular cleaning of the water runways. Since the possible sources of cooling water might be raw river water (rarely), well water, mains supply or recirculated factory water, regular samples for microbiological examination should be taken at relevant points such as retort sprays, cooling tunnels or at the inflow points in tank coolers. Plate counts at 25°C can give a good general picture of microbial load. In sampling where the water is chlorinated, it is important to neutralize the chlorine by the addition of sodium thiosulphate before analysis.

### Soft drinks

Ascosporogenous and Ascomycetous yeasts in fruit juice originate from the fruit skin and contaminate the processing equipment. Examples of such yeasts are *Saccharomyces cerevisiae*, *Saccharomyces uvarum* and *Saccharomyces bailii*. The filling and closure of bottles of fruit juice is not usually an aseptic process but should take place at about 80°C, otherwise viable contaminating yeasts, bacteria and moulds may enter and spoil the product. Sometimes if yeast is present in canned soft drinks it is an indication of insufficient plant sanitation. In fact, research has shown that the lower the efficiency of cleaning and disinfection of the plant during the production of fruit juice, fruit juice concentrate and soft drinks, the higher the chance that the micro-organisms will adapt to environmental factors such as the presence of anti-microbials in the cleaning material, and heat. Thus yeast may be disseminated through the circulating water and, despite the presence of anti-microbial materials, may cause problems of product contamination.

### Beer

Probably the most common microbial contaminant in English breweries is *Hafnia protea*, followed by *Hafnia alvea*, enterobacter species, Klebsiella species and *Citrobacter freundii* species.

All these species may be spread round the factory through the water supply and, if they contaminate the wort, they grow and produce smokey and phenolic flavours which are highly undesirable in the beer.

### Dairying

In dairying, as in brewing, when the raw product (in this case milk) is incubated while a starter culture is allowed to grow, it is vital that the product is not contaminated with any unwanted organism which can also grow while the fermentation occurs. An example of this happening is given on page 167. Contaminants and bacteriophage can be spread through the use of contaminated water in the plant.

### Other food processes

The examples already given (poultry, meat, canning, soft drinks, beer and dairying) illustrate the importance of using water of the highest microbiological quality at all stages in food processing. Water is used in many ways – as a blanching agent in the processing of vegetables, to wash fruit, to carry away grit from washed dried fruit and so on. In all these cases high water quality is all important. Where water is converted to ice, it is just as important because the contaminating organisms from the ice could survive in the food. Ice in contact with food should be of as high a microbiological quality as water in contact with food.

## Outgoing water and wastes

### Sewage

Sewage from factories and canteens is disposed of separately from the food waste and food polluted water, because of the need to guarantee that human intestinal pathogens are contained. Sewage is therefore discharged into the municipal system.

### Food polluted water and food wastes

Untreated process water which is discharged from food premises is liable to contain a high organic content and a high concentration of micro-organisms. These organisms would use up the dissolved oxygen in the receiving river or sea water and turn it stagnant so that it would not be suitable for fish; it would smell; it would become cloudy and would become unsuitable for use as a source of drinking water. The more organic material contained in the discharged effluent, the more oxygen in the receiving water would be used up. Thus a measure of the potential of water to go stagnant is called its 'biological oxygen demand' (BOD).

As mentioned at the beginning of this chapter, local authorities now require food factories to discharge their effluents at such a quality that the BOD of the receiving water is kept low and the water is not polluted to the point of stagnancy. Pollution control of the inland waters in the UK really got under way in the 1950s. The Rivers (Prevention of Pollution) Act, 1951 made it an offence to cause polluting matter to be discharged into any stream. Since that time, environmental improvement requirements, together with increased pressure on available resources, have made control even more important. The level of treatment required to prevent pollution depends on the organic content of the effluent.

It also costs money to treat effluent, and treatment was for this reason once resisted by food companies. The options available are either to use municipal facilities if the volume, strength and flow rate are acceptable to the authorities. To achieve acceptable levels balance tanks at the plant may be all that is required. The alternative is to install treatment plant on site at the factory.

There can be advantage to food manufacturers in treating their effluent, because food and other materials can be recovered from the effluent which otherwise would be wasted. For example, creamery effluent can contain a high fat content derived from milk, cream, butter or cheese manufacture, which if recovered may be sold for other uses.

In confectionery manufacture, waste water containing carbohydrates can be used as a feed stock for yeast fermentation plants, the yeast itself being separated and sold as a product for animal feed. Again, in the production of potato starch (which uses very high levels of water per tonne of potatoes), the untreated effluent contains very high levels of protein and starch and other insoluble material. In Holland this has polluted the canals for over a century, but environmental pressure has now demanded the treatment of such water. The food companies have responded to the demand and now recycle treated water for their own use, and the protein and starch they recover are converted into animal feeds.

In meat, poultry and fish processing, the waste waters contain sufficient protein and fats to be worth recovering from the water, and the recovered products can be used as additives in animal feeds. The efficient recovery of protein from factory effluents can reduce the BOD by about 75 per cent.

The treatment necessary can involve several stages – screening to remove solids, fat separation, aeration, filtration combined with settling, further filtration and possibly an activated sludge process. Filtrates or supernatants may be subjected to further purification by ultra-filtration and/or reverse osmosis to remove valuable dissolved organic materials such as proteins or lactose.

# Chapter 12

# Food fermentations

## Respiration and fermentation

In order to live, all organisms require energy. The process by which energy is released from the food materials – such as sugars – which the organism has taken in, is called 'respiration', and it may occur in the presence or absence of oxygen. The most commonly used substrate for respiration is the sugar glucose.

When oxygen is used the process is called 'aerobic respiration', and can be summarized:

sugar + oxygen → $CO_2$ + water +
            a large amount of energy

Anaerobic respiration or fermentation does not use oxygen and can be summarized:

sugar → $CO_2$ + alcohol (or other substance) +
        a small amount of energy.

The two formulae above represent processes which are really a chain of reactions taking place inside individual cells, each step of which is controlled by *enzymes*.

Enzymes are produced inside cells, and each has the role of aiding a particular reaction, often increasing its speed, yet at the end of the reaction the enzyme is unchanged and available for further activity.

The food upon which the micro-organism feeds is outside the cells, and before it can be used must pass inside the cell across the cell membrane. Many substances can enter the microbial cell with ease; other substances such as starch, fats and cellulose cannot because their molecules are too complex. In the case of cells

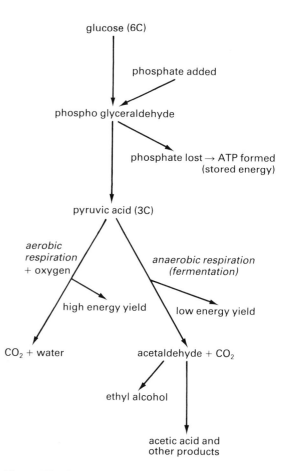

Figure 75  *Summary of respiration*

which use complex substances in their metabolism, enzymes pass out through the cell membrane and act on and break down the large molecules into smaller units which then pass into the cell.

Respiration involves the break down of glucose – a compound which has six carbon atoms – by a series of enzyme controlled steps, into two molecules of pyruvic acid – a compound which has three carbon atoms:

$$glucose\ (6C)$$
$$\downarrow$$
$$pyruvic\ acid + pyruvic\ acid$$
$$(3C)\qquad\qquad (3C)$$

Aerobic and anaerobic respiration differ from this point (see Figure 75). In *aerobic* respiration pyruvic acid is broken down by a large number of steps into carbon dioxide and water – a process which yields a lot of energy in a chemical form available for use within the cell. In *anaerobic* respiration far fewer steps are involved and the pyruvic acid is only incompletely broken down, yielding the end products carbon

dioxide, alcohol and lactic acid or other organic acid together with a small amount of energy (see Figure 76).

Micro-organisms are divided into groups according to their type of respiratory action. Strict *aerobes* require oxygen and cannot survive without it; strict *anaerobes* cannot respire in the presence of oxygen; and *facultative* organisms tend to respire aerobically under aerobic conditions and anaerobically under anaerobic conditions, being able to shift from one type of respiration to the other. Such a change can be induced in a culture respiring anaerobically by the sudden introduction of oxygen – an effect known as the Pasteur effect.

However not all organisms are so adaptable – for example the lactic acid producing organisms remain fermentative even under aerobic conditions.

The end products of respiration are released into the environment in which the organism lives, and if they are allowed to accumulate they tend to have an inhibitory effect on the growth of the cells which produced them. The fact that

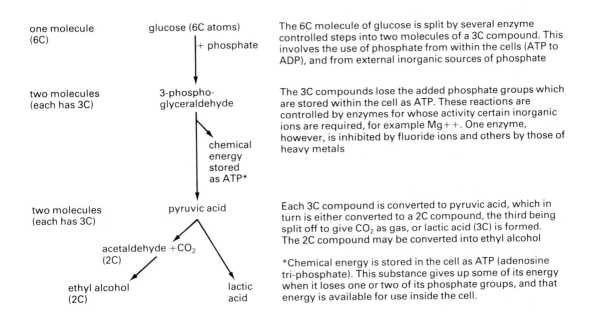

Figure 76   *Summary of fermentation*

the end products of respiration do accumulate has been exploited by man in the production of many types of food.

It is the anaerobic or fermentative type of respiration which is primarily utilized in the production of foods. When the end products are mainly alcohols, alcoholic drinks can be made – on a world-wide scale and throughout history, beverages such as beers, wines and saké have been produced. When the end products are primarily acids, other types of foods are produced; for example milk can be converted by fermentation into yoghurt, cheese, fermented drinks and other foods; vegetables are fermented to produce pickles and sauerkraut; the fermentative powers of yeasts are used in the leavening of bread.

When a culture of organisms is actively fermenting a substrate, its growth rate, that is, the rate at which the total cell number increases, slows down. Conversely, when active growth of a culture is taking place, fermentation is much reduced. In other words, growth and fermentation proceed at each others expense. So in the manufacture of foods dependent on microbial activity, the environmental conditions have to be carefully controlled to ensure that the desired level of cell increase or fermentation are being achieved.

## Foods produced as a result of alcoholic fermentations

### Bread

The organism which is primarily concerned in bread production is 'baker's' yeast – *Saccharomyces cerevisiae* – which is added to leaven the dough. By the evolution of carbon dioxide during the fermentation of carbohydrates in the dough, the dough volume is increased and at the same time flavour is imparted.

### Yeast fermentation in bread

In the dough the small amounts of glucose, fructose and sucrose from the flour, together with any added sugar, are fermented initially by the yeast.

The predominant carbohydrate, flour starch, is converted into maltose by the flour enzymes, α and β amylases, which are activated in the presence of water. This is known as the diastatic activity of the flour, and it may be aided by the addition of bacterial diastase which is readily available on the market. The yeast cells secrete the enzyme, maltase, which converts the maltose into glucose and this can then be taken in by the yeast cells. So the fermentation proceeds in two steps:

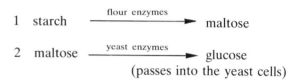

1   starch   $\xrightarrow{\text{flour enzymes}}$   maltose

2   maltose   $\xrightarrow{\text{yeast enzymes}}$   glucose
                                          (passes into the yeast cells)

The glucose so produced is fermented, with the result that the end products, carbon dioxide and alcohol, are formed. At the same time the flour protein, gluten, matures and becomes elastic and springy and capable of retaining the carbon dioxide evolved by the yeast. The conditioning of the gluten is due to the proteolytic activity of the flour and yeast enzymes, together with the physical kneading which the dough has received. In modern bread making on an industrial scale, the gluten may be developed very rapidly by intense mechanical mixing – such as in the Chorleywood Bread Process, or by the addition of cysteine. In these cases yeast strains which ferment carbohydrate rapidly must be used, otherwise the desired dough volume would not be achieved.

Thus it is important to encourage the production of $CO_2$ gas and to discourage the growth of the yeast cells – and this is achieved by manipulating the environmental conditions to favour fermentation.

The *water* requirement of the yeast cells is met in the moist dough. The requirement for *nitrogenous salts* and other inorganic ions can be met by the sparing addition to the mix of ammonium salts and calcium phosphate. These salts, together with added sodium chloride and sugar, can by osmotic effects, damage cells and deprive them of water if their concentrations are too high. In high sugar bun doughs where initial sucrose concentration can be up to 35 per cent, yeast activity may be reduced by up to 70 per cent and extra yeast must be added to compensate for the loss of activity.

A most important physical quality is the *temperature* at which the fermentation is allowed to proceed – a low temperature restricts gas production; high temperatures cause too much gas to be evolved and hence too great a volume is achieved before the gluten is adequately matured, and it fails to hold the gas. The temperature used is commonly between 25°C and 35°C, but is frequently adjusted to 26–28°C, a value well below the optimum (40°C) for yeast fermentation. The temperature which is selected is one which is commensurate with the rate of dough ripening – thus in bulk fermentation processes the temperature will be about 25°C, but in mechanical development processes fermentation and gas evolution must be rapid and a temperature of around 35°C is used.

The pH of freshly mixed dough is approximately 6.0 but falls as a result of the activity of the yeast and other organisms present to about 4.5 and is buffered at this value by the action of the flour. The yeast can ferment the sugars present at this pH value, but if acids are added as mould and rope inhibitors, the yeast is inhibited and there is loss of activity – compensated for by the addition of extra yeast.

Baking destroys the yeast cells – in fact cell destruction begins at 44°C.

It takes a while for the oven heat to penetrate the loaf, so yeast cells in the outer regions are destroyed first, but by the time it is baked yeast in the centre of the loaf is destroyed.

So, in bread-making the yeast is provided with environmental conditions which encourage fermentation and the production of carbon dioxide – the rate of evolution being controlled by adjustment of the environmental conditions, to ensure that the gas is trapped and held by the developing gluten to give a good loaf structure.

Bread is produced in this country in vast quantities on an industrial scale. Large quantities of yeast are required and this too is grown on an industrial scale.

*Industrial production of yeast*

The conditions under which yeast is produced on a large scale are those which encourage cell multiplication (growth) and discourage fermentation.

The medium which is used is either sterile beet molasses supplemented with the vitamin, biotin, or cane molasses, diluted to contain about 10 per cent sugar. To this are added ammonium salts, which provide nitrogen for protein synthesis, and other inorganic nutrients – such as phosphorus as superphosphate and magnesium sulphate. At this stage too, antifoam agents are added – which will reduce the foam formed as growth proceeds. The pH is lowered to 4.5 either by the old fashioned method of adding lactic acid bacteria and allowing the pH to drop as a result of their activity, or the newer methods of adding dilute mineral acid or ammonium sulphate. If the pH drifts too much on the acid side, the pH is raised by the addition of ammonium hydroxide. This pH – 4.5 – is effective in reducing the growth of bacterial contaminants.

The strain of yeast is carefully selected and cultured from a single cell in the laboratory and forms the 'seed' culture. This is used at the rate of 1.5 to 2.5 kg per 500 litres of medium to inoculate the large vats. The vats are well aerated – which discourages fermentative activity – by warm filtered air at 25–30°C for a period of about 12 hours, during which time the cell number increases five fold.

The cell-containing liquor is pumped from the vats and centrifuged to separate the cells from the liquor. The dense cell suspension is again washed and re-centrifuged and the cell cream is cooled and passed through filter drums to remove the excess moisture. The compact mass of cells passes to a cutting machine which divides it into 500 gm blocks which are wrapped and refrigerated. The blocks are stored at 1.5 – 7°C in a refrigerator in a chill room, and will remain in good condition for about 4 weeks. If stored at 10°C or above the yeast will deteriorate rapidly due to its metabolizing carbohydrate.

Active dried yeast is produced in the same way as the compressed yeast except that at the last stage the cream of yeast cells is pressed through a perforated plate and the threads of yeast which emerge are dried in a warm air current, dropping to the bottom of the container as granules, their moisture content reduced to about 8 per cent. The storage life of granules in airtight containers is several months. The active dried yeast can be reconstituted by soaking in water at 38 to 43°C, but the activity is below that of compressed yeast.

### Beer

Beer and other similar beverages are made from malted barley and water, and flavoured with the female flower of the hop plant.

The first stage in the production of beer is *malting* the barley. To do this the barley seeds are steeped two to four times in tanks of water at 10 to 15°C, a process taking up to 60 hours. The grain takes up to about 42 per cent moisture and is then put on to floors to a depth of about 100 mm, or into saladin boxes or drums where germination takes place.

The rate of germination may be hastened by the addition of the growth stimulating substance, gibberellic acid, at about 1 ppm. Then, at a time judged by the length of the acrospire and rootlets, the green malt is dried carefully (kilned), and then ground – the product being known as the *grist*.

The brewing process begins with *mashing*. During this 90 minute period the grist is mixed with water at approximately 70°C and the enzymes derived from the barley become very active and convert the starch into maltose. The *wort*, so produced, is then boiled in order to end the activity of the $\alpha$ and $\beta$ amylases, to sterilize it and to coagulate the residual proteins. The hot wort, to which have been added the flavouring hops or hop extracts, is cooled and aerated and then the yeast added, and the wort is said to be *pitched*.

The yeast is pitched into aerated wort in order to stimulate yeast cell growth, increase in yeast cell number and to encourage the synthesis of enzymes which will allow maltose to be absorbed and metabolized. The conditions

gradually change and become anaerobic and alcoholic fermentation begins and the pH drops to around 5.2.

The pitching yeast is often not a pure culture but is passed from brew to brew for years or even decades. This practice still continues but is being replaced gradually by use of a pure culture of pitching yeast produced in the brewery's laboratory. The yeast becomes contaminated by wild yeasts, acetic and lactic acid bacteria and other organisms as it is pitched from brew to brew, and so is discarded after about ten generations and replaced with new pitching yeast grown from a single cell strain. The flavouring hops possess antiseptic properties, which together with the lower pH and the presence of the alcohol help to inhibit the growth of undesirable bacteria, but use of pure culture pitching yeast means that the fermentation is more predictable, resulting in a predictable quality of beer.

In batch beer-making strains of *Saccharomyces cerevisiae* – a 'top' yeast – are used, which in the vigorous fermentation of the carbohydrate are carried to the surface of the beer on bubbles of carbon dioxide, and form a yeasty head which can be skimmed off; in lager-making, strains of *Saccharomyces carlsbergensis* – 'bottom' yeast – are used which do not rise to the surface of the lager but settle at the bottom of the fermentation vessel. The difference between 'bottom' and 'top' yeasts is becoming less significant than it has been in the past. In the development of continuous beer making processes, stirring is employed to keep the yeast uniformly suspended, and equal success has been achieved with either 'top' or 'bottom' yeasts. However, in batch processes, these flocculating qualities of the strain of yeast used must be carefully monitored, because if there is an alteration from the normal pattern, due for example to the growth of contaminating organisms, separation of cells from product cannot follow the normal pattern and a yeasty product could result.

The carbon dioxide evolved during fermentation is led away and can be stored and used in the carbonation of beers, or in the manufacture of 'dry ice', or, as in some breweries, not recovered at all.

The fermentation of the glucose derived from the maltose is allowed to run until it is all used up – the beer at that time having an alcohol content of 2.5 to 10 per cent. Then after the yeast cells have been separated from the beer, it is run into storage tanks and either put into casks, racked, or it is filtered and bottled and then pasteurized at 68°C for a few seconds or at 60°C for 20 minutes.

Spoilage of beer is primarily due to wild yeasts which have fermentation patterns differing from that of the pitching yeast, and may cause problems in the clarification of beer. Yeasts of the genera *Pichia* and *Hansenula* may form pellicles, and certain strains of *Saccharomyces* impart unpleasant flavours.

Beer may be acetified by the growth of bacteria of the genera *Acetobacter* and *Acetomonas*. The danger of this arises particularly in cask beer when air has entered the cask to replace beer which has been drawn off. The same organisms may also, by capsular secretions, cause 'rope'. Lactic acid bacteria – both rods and cocci – can spoil beer under anaerobic conditions. They are resistant to the antiseptic qualities of hops and cause turbidity, off flavours, acid and sometimes ropiness.

## Wine

Wine results from the alcoholic fermentation by yeasts of the sugars glucose and fructose contained in grapes and other fruits. Wine yeasts can grow well in the highly acid conditions (pH 3–4) in grape juice and can resist 10 per cent or more alcohol, and resist the sulphur dioxide which is added to repress spoilage bacteria.

When making *red wines* the grapes are crushed, and water and sugar added if it is considered necessary. Sulphur dioxide is added as bisulphite to a level of approximately 100 ppm to suppress the growth of spoilage bacteria and yeasts.

The fermentation is allowed to begin – the starter organisms are sometimes the natural yeasts found on the skin of the grapes. More commonly these days a specially grown culture of *Saccharomyces cerevisiae var. ellipsoideus* is added to the non-sterile grape must. At first, conditions favour the growth of the yeast, but they soon become anaerobic and encourage fermentation.

The temperature of the *must* (crude fruit pulp) is carefully controlled at 24 to 27°C for a period of 3 to 5 days. If it is not so controlled spoilage bacteria or yeasts could grow and become predominant, in addition to which, the fermenting yeast may be killed. As the alcohol content steadily rises the red alcohol soluble pigment in the grape skins is extracted giving colour to the wine. Gas produced causes the pulp to float to the surface, and if this is not regularly broken up, it can be the site in which rapid growth of spoilage organisms occurs. The fermentation period is several weeks, after which time the alcohol content can be as high as 19 per cent, but typically is 11 to 17 per cent, the yeast normally being inhibited at alcohol concentrations beyond 18 per cent. The wine is drawn off from the pulp and aged in vats when the flavour develops.

*White wines* are made in a similar way except that the skins and pulp are removed from the juice before fermentation begins and the temperature of fermentation at 10 to 15°C is lower than that for red wine. In sweet wines a proportion of the sugar present remains unfermented.

Wines can be pasteurized, a process which is primarily to clarify the wine by precipitation of proteins, but also to destroy a proportion of spoilage organisms. Wild yeasts may cause the development of pellicles; acetic acid forming bacteria of the genus *Acetobacter* may acidify the wine, as may lactobacilli by the production of lactic acid. The presence of acetic acid at 0.1 per cent or more inhibits yeast growth and fermentation. Wine may become slimy or ropy due to the growth of members of the genus *Leuconostoc* and other organisms.

## Vinegar

The members of the genus *Acetobacter* are used in the production of vinegar by the conversion of ethyl alcohol to acetic acid. They are allowed to grow aerobically on the surface of wine to produce wine vinegar or to acetify unhopped beer to produce malt vinegar.

Its production may be summarized:

$$\text{glucose} \xrightarrow{\text{(fermented by yeasts)}} \text{alcohol} + CO_2$$

$$\text{alcohol} \xrightarrow{\substack{\text{(oxidation by acetic} \\ \text{acid organisms)}}} \text{acetic acid} + \text{water}$$

## Foods produced as a result of acid fermentations

Lactose is a disaccharide, which occurs in milk and which can be split into its constituent monosaccharides by microbial enzymes. The resulting galactose and glucose are then fermented to lactic acid:

$$\text{lactose} \longrightarrow \text{glucose} + \text{galactose} \longrightarrow \text{lactic acid}$$

The homofermentative lactic acid streptococci and the lactobacilli are responsible for this type of reaction where the end product is predominantly lactic acid.

Heterofermentative organisms, for example, members of the genus *Propionibacterium*, may metabolize glucose beyond lactic acid to propionic acid, acetic acid and gas ($CO_2$), or possibly to a more complex range of end products – acetic acid, butyric acid, ethyl alcohol, butyl alcohol and acetone, and the gases carbon dioxide and hydrogen. Sometimes bacteria ferment carbohydrates to produce ethyl alcohol as the predominant end product.

$$\text{glucose} \longrightarrow \text{lactic acid} \longrightarrow \substack{\text{propionic acid} + CO_2 \\ \text{acetic acid} \\ \text{and other acids}}$$

The ability of micro-organisms to ferment glucose to lactic acid and other end products is used in the production of a wide variety of foods: fermented milk products such as cheese, butter, yoghurt and pickled vegetables.

### Cheese

To milk which has been pasteurized and cooled to about 29°C is added a 'starter' culture of lactic streptococci. These organisms grow and ferment the milk lactose to lactic acid, so lowering the pH of the milk from pH 6.6. Then rennin – a proteolytic enzyme prepared from the stomach of calves – is added. In the acid conditions provided by the bacteria the rennin rapidly coagulates the milk protein, casein. This clot, known as a 'coagulum' is cut into very small pieces by special knives in order to aid the expression of the whey – the fluid which remains. The coagulum is gathered together and, after a series of pressings to give it a firmer texture, is finally cut into very small pieces and salted. The salted cheese is then packed into moulds, pressed and stored. The cheese is tended carefully as it matures because it is at this time that the flavour develops, yet at the same time it is subject to spoilage. The flavour development (ripening) is due to the slow growth of micro-organisms within the cheese and these produce diacetyl, carbonyls, amino acids, amines and other aromatic substances from the slow hydrolysis of fats and proteins.

The English cheeses of the Cheddar, Cheshire and Wensleydale types are 'bacterially ripened' and do not require the addition of any organisms to aid flavour development. Other cheeses produced in basically the same way are allowed, during the maturation period to develop mould – particularly of the *Penicillium* species – on their outer surfaces, as is the case with Brie and Camembert; or mould growth is encouraged within the cheese by piercing it with long rods as in Stilton and Danish Blue. These cheeses are 'mould ripened'. The 'eyes' (holes) in cheeses such as Gouda, Gruyere and Emmental are produced as a result of the growth of organisms which produce carbon dioxide in large volumes, and the gas is trapped within the cheese structure and causes the development of holes. Organisms responsible for this are notably those belonging to the genus *Propionibacterium*.

### Butter

Butter is produced from the separated fats of

milk. The cream is pasteurized and then the starter culture is added, which in its growth over 1 to 2 days at 15 to 21°C causes the development of acidity and flavour, that is ripening. Acid is developed by the vigorous growth of *Streptococcus lactis* and *Streptococcus cremoris*, and aroma by the slower growth of organisms which utilize the citric acid in the cream – *Streptococcus dextranicus* and *Streptococcus citrovorus*. The cream must be stirred regularly to avoid the growth of spoilage organisms. After ripening, the cream is churned until a phase inversion occurs and it becomes converted from a fat-in-water to a water-in-fat emulsion, a process aided by the development of acidity. The butter is removed and, after washing, can be salted. It is then worked to exclude excess moisture, shaped and packaged. Butter which has not been ripened has better keeping qualities.

### Yoghurt

Yoghurt is made from pasteurized milk to which a starter culture is added.

The starter used is a mixture of *Streptococcus thermophilus* and *Lactobacillus bulgaricus*. During an incubation period at about 35°C these organisms ferment the milk lactose and produce acid. Under acid conditions the casein coagulates and the yoghurt is produced.

Similar foods are produced in other countries in which cows', goats', mares' or other milk is used to produce fermented products. The nature of the product may be liquid or solid, acid, acid and fizzy (due to gas production) or alcoholic. Examples of these products are kefir, koumiss, cultured butter milk, sour cream, acidophilus milk and there are many others as well.

### Pickled vegetable products

In addition to fermented milk products some vegetables are fermented.

Many vegetables which would spoil rapidly can be preserved as pickles in the presence of acid. Often vinegar is used for this purpose, but on some occasions the acid is allowed to develop naturally as a result of the growth of lactic acid producing organisms. Sauerkraut (pickled cabbage) and olives are two such examples, while in the preparation of onions, cauliflower, walnuts, etc., the development of lactic acid prior to the addition of vinegar is of importance.

In various parts of the world many foods are fermented to produce locally favoured products. For example in the East, rice is steamed and the mould *Aspergillus oryzae* is cultured on it, followed by a yeast fermentation in which up to 20 per cent alcohol is produced. This is the basis of making the drink saké. In Africa many products involve lactic fermentations of which Busa is notable.

All organisms respire in order to produce energy for their own use. Anaerobic respiration – *fermentation* – is exploited by man to produce several types of food. Alcoholic fermentations are made use of in the production of bread, beer, wine; while acidic ones are used in the production of cheese, butter, yoghurt, pickles and other fermented vegetable products.

# Appendix 1   Legislation

**A summary of legislation relating to food hygiene, the microbiological quality of food and water in the UK**

In this Appendix relevant legislation is summarized. Parts of the regulations are quoted, some parts are précised and others are omitted altogether. Anyone wishing to know the exact word of law should refer to the Statutory Instrument concerned, a copy of which can be obtained from Her Majesty's Stationery Office.

## Introduction

The basic food legislation in the UK is contained in the following:

1  The Food and Drugs Act, 1955 and its amendments, The Food and Drugs (Control of Food Premises) Act, 1976 and The Food and Drugs (Amendment) Act, 1981. The Act applies primarily to England and Wales; it does not apply to Scotland and only in certain instances to Northern Ireland
2  The Food and Drugs (Scotland) Act, 1956
3  The Food and Drugs (Northern Ireland) Act, 1958

Each act is divided into parts under which regulations can be made. The parts into which the 1955 Act is divided are:

1  General provisions as to food and drugs
2  Milk, dairies and cream substitutes
3  Provision and regulation of markets
4  Slaughterhouses and knackers' yards; cold air stores

5  Administration, enforcement and legal proceedings
6  Miscellaneous and general

The 1956 and 1958 Acts are also divided into parts in a similar, but not identical manner.

In general terms this basic food legislation makes it an offence to render food, which is going to be offered for sale for human consumption, injurious to health. This condition may result from the addition of substances, the subtraction of constituents or the manner in which the food is treated and handled.

General protection is given to the purchaser of food (or drugs) which 'is not of the nature, substance, or quality demanded by the purchaser'.

The composition and labelling, and the manner in which food is produced, are all governed by the Act. Labelling food in a manner which falsely describes the contained food is an offence. To ensure that the standards required are maintained the Regulations made under the Act provide local authorities with the right to take samples, to have them analysed and to prosecute where they consider it to be necesary. The Act contains powers to make regulations concerned with food hygiene; and requires the registration of premises, and the licensing of vehicles in connection with the sale of food.

## Statutory Instruments

In England, Wales and Scotland sets of Regulations made under various Acts are known as

Statutory Instruments (SI) and each SI is given a number. In Northern Ireland the 'sets' of Regulations are known as Statutory Rules and Orders (SR and O) and these too are numbered. New SI and SR and O are always being made to keep up with modern practices, so previous Regulations may be amended by additional Regulations or may be entirely replaced.

### Keeping up to date

Any summary of 'current' legislation soon goes out of date. Her Majesty's Stationery Office issues a list every month, which is summarized at the end of every year entitled:

'List of SI for the month of ... 19. .' and
'List of SI for the year 19. .'

These two publications provide outline details of all the legislation which has passed through Parliament.

Similar lists of SR and O which relate to Northern Ireland are published by HMSO (Belfast).

In addition the HMSO issues every year, an 'Index to Government Orders' which under alphabetical headings lists the Orders which are in force on the last day of the year.

Copies of Regulations can be purchased from HMSO or reference can be made to *Butterworths' Law of Food and Drugs* (which supercedes Bell and O'Keefe's *Sale of Food and Drugs*) which gives full details, together with notes and comments, but it does not consider SR and O.

EEC Regulations and Directives are published in the *Official Journal* of the EEC available from HMSO. Otherwise, the *British Food Journal*, published six times a year, has a very useful section on food legislation, both current and impending.

## Summary of regulations

The Regulations summarized in this Appendix are those which have direct bearing on the microbiological quality of food, food hygiene and water used in food processing and handling.

1 Food hygiene
   (a) General regulations concerning food hygiene
   (b) Markets, stalls and delivery vehicles
   (c) Docks, carriers, etc.
   (d) Slaughterhouses
2 Protection of the microbiological quality of food
   (a) Milk
      (i) production
      (ii) designations of milk
      (iii) addition of preservatives to milk
   (b) Ice-cream
   (c) Liquid egg
3 Control of addition of substances to foods
   (a) Addition of preservatives
      (i) general
      (ii) milk
   (b) Addition of other substances
      (i) meat
4 Control of techniques of processing
   (a) Irradiation
   (b) Sterilization
5 Other regulations
   (a) Meat inspection
   (b) Imported food
   (c) Shellfish
   (d) Poultry
   (e) International carriage
6 Water
7 Waste disposal
8 Food hygiene codes of practice
9 Food legislation in the EEC

## Food hygiene

Food hygiene regulations are concerned with food which is or is intended to be for sale for human consumption. They aim to secure sanitary and clean conditions and practices while food is exposed for sale, or during its importation, transport, storage, packing, wrapping service or delivery. The object is to protect public health.

### General regulations concerning food hygiene

The regulations in force at the moment are:

*England and Wales* The Food Hygiene (General) Regulations, 1970 (SI 1970, No. 1172).

*Scotland* The Food Hygiene (Scotland) Regulations, 1959 (SI 1959, No. 413). Subsequent amendments: SI 1959, No. 1153; SI 1961, No. 622; SI 1966, No. 967; SI 1978, No. 173.

*Northern Ireland* The Food Hygiene (General) Regulations (Northern Ireland), 1964 (SR and O 1964, No. 129).

The Regulations for the whole of the UK are substantially the same, but where important differences exist they are detailed in the text below which summarizes The Food Hygiene (General) Regulations, 1970 (SI 1970, No. 1172).

The Regulations apply to the supply of food in the course of business from *fixed* premises and home-based ships,* that is, any trade or business for the purposes of which any person engages in handling food: for example, canteen, club, school, hospital, institution, any undertaking carried on by public and local authorities, retail sale of food, etc.

Regulations for *unfixed* premises (markets, stalls and delivery vehicles) are considered later on page 188. Separate Regulations apply to docks and carriers (see page 189).

### Protection of food from risk of contamination

The Regulations require that food is protected from the risk of its contamination by dirt, other foods or from any source which might contain pathogenic organisms.

There must be no risk of contamination of the food due to the siting or construction of the premises – they must not be insanitary.

Any equipment used in handling the food, and wrappers and containers of food, must be kept clean, and all equipment with which food comes into contact must be made of materials which can be thoroughly cleaned and prevent as far as reasonably practicable any matter being absorbed by them.

Unprotected food must not be placed within 18 inches (24 inches in NI) of the ground, and it must be kept away from animals and other sources of infection, such as food unfit for human consumption, and adequately protected while offered for sale. The food handler must not use any wrapping paper liable to contaminate the food – in particular, newspaper – or allow it to come into contact with open food.

### Personal cleanliness

Since the human body is a source of organisms and the means by which pathogens can be transferred from one food to another, the law requires that personnel who work with open food should have a very high standard of personal cleanliness.

The handler should have a high standard of personal hygiene, keeping all parts of his person which are liable to come into contact with food as clean as possible and wear clean protective overalls. Every person who carries meat which is open food and which is liable to come into contact with his head and neck must wear a clean and washable neck and head covering. Clothing and footwear which is not worn in working hours by all persons engaged in handling food is to be kept in cupboard or locker accommodation provided for this purpose.

Any open cut or abrasion on any exposed part of the food handler's person should be covered with a suitable clean waterproof dressing. First aid materials are to be provided in the premises in a readily accessible position. The first aid box must contain bandages, dressings (including waterproof ones) and antiseptic.

The food handler must refrain from spitting and the use of tobacco, other smoking mixture or snuff while handling open food, or while he is present in any room in which there is open food.

If a person engaged in the handling of food is suffering from, or is a carrier of, typhoid fever, paratyphoid fever or any Salmonella infection, amoebic or bacillary dysentery, or Staphylococcal infection liable to be the cause of food

---

* See also The Food Hygiene (Ships) Regulations, 1979 (SI 1979, No. 27).

poisoning, he or she is to inform the person carrying on the food business who must immediately inform the appropriate Medical Officer of Health.

*Hand-washing, lavatory and cloakroom facilities*
The hands are one of the important means by which pathogenic micro-organisms are transferred directly from one food to another, from foods to surfaces and equipment, and from there to other foods. The human bowel can contain pathogens and in the use of the lavatory these organisms can be transferred to hands. For these reasons the law requires frequent and effective hand-washing between tasks and after the use of the lavatory.

Sanitary conveniences, well lit and ventilated, are to be kept clean and in working order, and so placed that no offensive odours can penetrate into any food room. No room which contains a sanitary convenience is to be used for handling open food or as a food room. There is to be fixed and maintained in a prominant position near every sanitary convenience a clearly legible notice requesting users to wash their hands after using the convenience, such as 'Now wash your hands'.

Sufficient wash basins for the use of all persons engaged in handling open food are to be provided. These are to be kept clean and in working order and provided with either hot and cold water or hot water at a suitably controlled temperature. At each basin there is to be an adequate supply of soap or other suitable detergent, nailbrushes and clean towels or other suitable drying facilities. The hand-washing basins are not to be used for any other purpose.

*Washing food and equipment*
Foods are often dirty and also are frequently sources of pathogenic organisms. Such food contaminates all that it touches – hands, chopping blocks, utensils, conveyor belts, knives, bowls, basins, etc. The Food Hygiene Regulations require therefore that facilities should be provided to wash foods; and also that equipment which comes into contact with food can be thoroughly cleaned and sterilized.

Sinks, or other suitable washing facility (such as a machine) with hot and cold water, must be provided for washing equipment. Some sinks need have only cold water provided, for example those which are used solely for washing fish, fruit or vegetables; and also, in conjunction with a suitable bactericidal agent, sinks used solely for washing drinking vessels or only ice-cream formers or servers. All sinks must be kept clean and in working order. The Scottish Regulations specify the manner in which a utensil must be cleaned after its use for the service of food to a consumer, and before its re-use: either wash in clean hot water, rinse in clean water at not less than 76·7°C; or wash in clean hot water, rinse in clean hot water containing an effective bactericidal agent, rinse with clean water; or wash in clean hot water containing an effective bactericidal agent, rinse in clean water, after which dry by evaporation, or with a clean cloth.

The Scottish regulations also specify the manner in which drinking vessels, other than those used for milk or other drinks containing more than 4 per cent fat, may be washed. They may be cleaned in machines provided that the vessel is washed in clean water containing a suitable quantity of bactericidal agent and afterwards rinsed with clean water, and allowed to dry or be dried with a clean cloth.

The Scottish Regulations also specify the manner in which equipment which has been used in the preparation or storage of meat or fish products, or with which bakers confectionery filling somes into contact, is to be cleaned: 'First such equipment must be cleaned – to remove traces of organic soil – and then at least once on every day on which it is used be treated with steam, or water, at a temperature not lower than 76·7°C, or washed with clean hot water containing an efficient bactericidal agent in suitable quantity, and thereafter rinsed with clean water. Tubs and containers used for brining meat are to be treated in the same way after each occasion on which the brine is removed'.

*Temperatures at which certain foods must be kept*
Some foods support the growth of food poison-

ing organisms and if they are held 'warm' for any length of time any contaminants which they contain will have the opportunity to multiply. The law seeks to avoid this happening.

All food consisting of meat, fish, gravy or imitation cream (or those containing any of them) or any egg or milk must be kept either at or above 62·7°C or below 10°C and not be allowed to remain within this range except as a step in preparation and kept within this range for the minimum practicable period. Food which is exposed for sale is exempted from this requirement – but in accordance with good practice it is advisable that food offered for sale should as far as is possible be held at the temperatures required in this Regulation.

Certain foods – which are not likely to support the *growth* of food poisoning organisms because they are too dry, or contain high levels of sugar or salt, or are otherwise preserved – are also exempted from this regulation: bakery products prior to baking, chocolate or sugar confectionery, canned and bottled foods, butter, margarine, lard and other fats, ice-cream to which the Ice Cream (Heat Treatment etc.) Regulations apply, cheese, uncooked bacon and ham, dry pasta, dry pudding mixes and other dry mixes, unskinned rabbits, hares, unplucked game and poultry.

The Scottish Regulations also specify conditions for reheating food, for the treatment of gelatine and bakers confectionary filling:

*Re-heating of food*: food which has been heated, and which is re-heated prior to sale or service, must be reheated to not less than 82°C.

*Gelatine* for use in the preparation of bakers confectionary filling, meat or fish products must immediately before use be brought to the boil, or brought to at least 71°C for 30 minutes. Any left over gelatine if not wasted must be quickly cooled under hygienic conditions and kept in a refrigerator or other cool place.

*Bakers confectionary filling* (cream, reconstituted, imitation or other cream substitute or any filling containing cream) must not be touched by hand in any process where it is reasonably practicable not to. Any person about to engage in using the filling shall wash his or her hands before using it – especially if he or she has to touch it.

*Premises*
The Regulations impose certain conditions regarding the structure and maintenance of premises with the object of ensuring that premises are as hygienic as possible.

Guidelines are laid for the design of soil drainage systems and the siting of cisterns for water supply. The Regulations also require that premises are kept in good repair. The walls, floor, windows, ceilings, woodwork and all other parts of every food room must be kept clean, in good order and condition to enable them to be effectively cleaned and as far as is reasonably practicable prevent any risk of infestation by rats, mice or insects.

Both outside the premises and within, adequate space, suitably sited must be provided for the temporary storage of refuse prior to disposal, and refuse of any sort must not be allowed to accumulate within a food room.

Food rooms must be well lit and ventilated, and not used as sleeping places.

### Markets, stalls and delivery vehicles
The Regulations concerning hygiene in markets, stalls and delivery vehicles in England and Wales are to be found in: The Food Hygiene (Markets, Stalls and Delivery Vehicles) Regulations, 1966 (SI 1966, No. 791 and amended by SI 1966, No. 1487).

The Regulations for the whole of the UK are substantially the same – but the notes below *summarize* the Regulations relating to England and Wales (SI 1966, No.791).

The standard of hygiene which the law requires at *unfixed* premises conducting food businesses is the same as that required for *fixed* premises – and as outlined under the general regulations. Persons engaged in handling open food must observe the same standards of personal cleanliness and hygiene.

In addition a stall or delivery vehicle must have displayed on it the name and address of the person carrying on that business, and any other address at which it is normally garaged.

There must be suitable and sufficient space for the separation of unsound food, and for the disposal of waste, and sufficient covered receptacles for waste trimmings, refuse and rubbish.

Except where only raw vegetables are sold, stalls selling open food for human consumption must be covered at the sides and back to prevent mud, dust, dirt, filth, etc., from being deposited on and contaminating the food.

The hygiene of the *transport of meat* is also controlled by these regulations. In Northern Ireland it is controlled by the Food Hygiene (General) Regulations, and in Scotland by The Food (Preparation and Distribution of Meat) (Scotland) Regulations, 1963 (SI, 1965 No.2001 and SI 1967, No.1507). Although differences exist between the requirements of these regulations, their object is the same: to protect open food (meat) while in transit from undue contamination. In England and Wales the requirements are:

'That the vehicle used for the transport of meat must be closed by canvas or other washable material, the floor must be impervious or fitted with moveable duckboards which are used in such a way as to prevent meat touching the floor, any equipment used for loading or unloading which comes into contact with meat must be kept clean and in good order to enable it to be thoroughly cleaned. Suitable impervious containers must be used to contain separately all offal. This does not include packaged and frozen offal, or giblets of game and poultry still attached to the carcase from which they have been removed. Exception is also made for uncleaned tripe, stomachs, intestines, feet – in a vehicle used solely for these types of offal; and also for unskinned or unscalded heads in a vehicle used solely for these types of offal'.

The Food and Drugs (Control of Food Premises) Act, 1976*, amends the Food and

Drugs Act, 1955, and contains powers to prohibit the sale of food where there is imminent risk of danger to health.

### Docks, carriers, etc.

Regulations have been made which are concerned to secure the hygienic handling of food at docks, warehouses, cold stores, carriers, premises and other special premises, such as slaughterhouses, specifically excluded from the general regulations.

In regard to docks, carriers, etc. the Regulations require the cleanliness of premises, vessels, vehicles used for the reception of food, and the cleanliness of workers and their clothing. If personnel suffer from or are carriers of infectious diseases this must be immediately reported in the same way as required in the general regulations. They contain requirements on construction, maintenance and repair of premises, the provision of sanitary conveniences, the provision of a water supply and washing facilities.

See The Food Hygiene (Docks, Carriers etc.) Regulations, 1960 SI No.1602, amended by SI 1962, No. 1287.

### Slaughterhouses

The Slaughter of Poultry Act, 1967 provides for the humane slaughter for commercial purposes of poultry (turkeys, domestic fowl, guinea fowl, ducks and geese). It makes provision for certain methods of slaughter; requires the registration of premises and gives powers of enforcement to local authorities. It applies to England, Wales and Scotland. It is amended by the Slaughterhouse Act, 1974. This consolidates certain enactments relating to slaughterhouses and knackers' yards and to the slaughter of animals. It requires licensing of slaughterhouses, slaughter areas and knackers' yards. It permits only certain methods of slaughter and seeks to ensure that humane conditions exist.

Under these two Acts, Regulations may be made including the hygiene regulations outlined below.

---

* The control of Food Premises (Scotland) Act, 1977 amends the 1956 Food and Drugs Act. Food Premises in Northern Ireland are regulated under SI 1979, No. 1710.

*Hygiene*

The Regulations are designed to ensure that slaughterhouse construction, operation and handling of meat are hygienic. They impose upon slaughterhouse operators and others, requirements as to construction, layout, drainage, equipment, maintenance, cleanliness, ventilation, lighting, water supply, management and personal hygiene.

The *layout* of a slaughterhouse must provide adequate space for all operations. For example, to permit clean operations to be separated from those liable to give rise to contamination. It must be such as to permit all operations to be carried out hygienically.

Many of the requirements are similar to those in the general regulations including the supply of first aid materials, notification of disease suffered or carried by personnel, personal cleanliness and the wearing of clean overalls, head covering and boots. Requirements specific to slaughterhouses are that the lairages for the animals must be kept clean, and so arranged that it is possible to separate diseased or suspected diseased animals from the others, and that only live animals may be brought into the slaughterhouse. The time period in lairage is restricted to a maximum of 72 hours prior to slaughter.

Separate receptacles for blood, garbage, offal and refuse must be supplied. Refuse is to be removed as often as necessary to prevent a nuisance and in any event once every two days, and the refuse receptacles are to be thoroughly cleaned before re-use. All fixtures, surfaces, walls, floors, etc. are to be kept clean allowing no accumulation of debris; and the entry of flies, rats, mice and insects must be prevented as far as possible.

Any person using any knife, scabbard, sharpening steel, chopper or saw in a slaughterhouse must ensure that it is thoroughly cleansed and disinfected in water at a temperature of not less than 82°C – frequently during the course of the day; immediately after any contact with meat known or suspected to be diseased; before re-use, after any break in work; at the end of each working day.

No person shall in a slaughterhouse wipe down any carcass or any offal. Every person who comes into a slaughterhouse from a knacker's yard must *before* handling any meat or blood intended for human consumption thoroughly wash all parts of his or her person that may come into contact with such meat or blood, and change into clean protective clothing. See The Slaughterhouse (Hygiene) Regulations, 1977 (SI 1977, No.1805); The Slaughterhouse (Hygiene) Scotland Regulations, 1978 (SI 1978, No. 1273); and The Slaughterhouse (Hygiene) Regulations (Northern Ireland), 1963 (SR and O 1963, No. 162). For general guidance see Circulars FSH 13/77 and FSH 5/78.

The Fresh Meat Export (Hygiene and Inspection) Regulations, 1981, SI No.454 are concerned with export slaughterhouses and lay down conditions for approval of premises, slaughter, dressing, cutting of meat, inspection, hygiene, health marking, certification, storage, wrapping, packing and transport of meat. They are designed to implement EEC Directive 64/433 on health problems.

Other existing Regulations still apply (such as The Slaughterhouses (Hygiene) Regulations, 1977 and The Food Hygiene (General) Regulations, 1970, but they amend parts of The Meat Inspection Regulations, 1963 (SI 1963, No. 1229).

## Protection of the microbiological quality of food

### *Milk*

*Production*

Milk is a food which when produced from a healthy cow is sterile. It obtains its flora of organisms from the teat ducts and udders, and subsequently from the environment—airborne dust, the milker's hands and the milk receptacles such as churns or tanks. The organisms which contaminate the milk may be pathogens or spoilage organisms. Either way their presence is to be avoided as far as possible, and the milk handled in a manner not likely to encourage the multiplication of those present. The legislation in the three regions of the UK differs but its objectives are the same. The general conditions under which milk is produced in England and Wales are governed by The Milk and Dairies (General) Regulations, 1959, SI No.277 and its amendments SI 1977, No. 171 and SI 1979, No. 1567.

An outline of the English Regulations is as follows. Dairy farms and farmers, and distributors of milk must be registered with the Ministry of Agriculture, Fisheries and Food. Dairy cattle must be inspected by veterinary surgeons at regular intervals to ensure that the cattle are healthy, and not suffering any disease. The milking houses are to be clean inside and outside, and the water which is supplied to the cows must be clean and wholesome. Before milking, the udders, teats and flank region of each cow must be cleaned, the hands of the milker must be clean, and the receptacle which receives the milk must be such that it can be covered. Milking must be done in good light. The milk must be cooled to a temperature of not more than 10°C after milking unless it is immediately heat-treated or sold to a consumer or distributor. A distributor is required to ensure that the milk is cooled to the same degree.

The conditions under which milk is handled and processed must be such that the risk of its contamination is minimized; it must be covered to protect it from flies, dirt, dust, etc., and WCs, cesspits and manure heaps must not be sited close to the dairy to reduce the risk from these sources. Personnel must be clean, and cover open cuts and abrasions with clean waterproof dressings, they must refrain from spitting and the use of tobacco and snuff while in the dairy. Milk which is being conveyed or stored must be protected from accidental contamination.

If any person who has access to milk or churns on registered premises, or if any member of his or her family is affected by a notifiable disease* the occupier of the registered premises must be notified, who must in turn notify the Medical Officer of Health in the district in which the premises are situated.

The Regulations outline how vessels, utensils and appliances are to be cleaned. Every vessel and its lid are to be thoroughly clean immediate-ly before use. Milk tankers, vessels or appliances after use are to be rinsed and washed with or without detergents, and before they are used again they are to be scalded with boiling water or steam or otherwise cleaned with an approved chemical agent. Acceptable chemical agents are set out in circulars FSH 2/73 and FSH 7/73 and include the following categories:

Sodium hypochlorite solutions
Detergent/sterilizers containing QACs
Detergent/sterilizers based on available chlorine
Solutions of QACs
Iodophors
Ampholytic surface active compound sterilizer
Sterilizer based on available chlorine

*Designations of milk*

In England and Wales the categories of milk which may be sold are 'untreated', 'pasteurized', 'sterilized' and 'ultra heat treated'. These are defined in The Milk (Special Designations), Regulations, 1977, SI 1977, No. 1033 and its amendments SI 1980, No. 1863.

Untreated milk is milk which has not been treated at any stage, and is produced from cows of an accredited herd. Such milk must pass the half-hour methylene blue test.

Pasteurized milk is milk which has either been retained at a temperature of not less than 62·8°C and not more than 65·6°C for at least 30 minutes and immediately cooled to a temperature of not more than 10°C (7·2°C in Scotland and Northern Ireland), or raised to a temperature of not less than 71·7°C for 15 seconds and cooled to a temperature of not more than 10°C (7·2°C in Scotland and Northern Ireland).

This milk has got to pass the phosphatase test (see footnote*on page 192), but if also tested with the MB test must satisfy that too.

---

* Notifiable disease means food poisoning, gastroenteritis and smallpox, cholera, diphtheria, membranous croup, erysipelas, scarlet fever, typhus, typhoid, enteric fever, relapsing fever, or puerperal fever. (Defined in the Public Health Act, 1936, and the Public Health (London) Act, 1936.)

*Sterilized milk* is milk which has been filtered or clarified, and homogenized, and then heated to a temperature of not less than 100°C for such a period that the milk will satisfy the turbidity test* as laid down in the regulations.

*Ultra heat treated milk* (UHT) is milk treated at a temperature of not less than 132·2°C for not less than 1 second and immediately put in sterile containers which are then sealed. Such milk must pass the 'colony count test'.*

In Scotland† four categories of untreated milk exist – none is heat treated at any stage, but they differ in the minimum microbiological quality which they are required to attain. They are summarized in Table 38.

In addition to the untreated categories, pasteurized, sterilized and UHT milk are designated and defined in the same way as in England and Wales – but the tests they are required to pass are not identical (see Table 39).

In Northern Ireland untreated milk is described as *farm bottled*. Such milk is to be cooled to a temperature not in excess of 7·2°C prior to sale, and on testing must contain no more than 50,000 bacteria/ml, and must pass the methylene blue reduction* test.

*Pasteurized milk*, defined in the same way as in the English Regulations, must pass the phosphatase test*, and contain no coliform in 1/10 ml. The Regulations which define the designations of milk in Northern Ireland are The Milk Regulations (Northern Ireland), 1963, SR and O No.44 and as amended by SR and O 1965, No.46.

All the Regulations require that the milk is sold in sealed containers and labelled according to its designated type. The tests and standards which the milks must pass are summarized in Table 39, but although the names of the tests for the different regions are the same, such as

'phosphatase test,' the standards and conditions of test are not necessarily exactly the same.

The Food and Drugs (Milk) Act, 1970, authorizes the treatment of milk by the application of steam. The Milk and Dairies (Semi-skimmed and Skimmed Milk) (Heat Treatment and Labelling) Regulations, SI 1973, No. 1064 define the methods for pasteurization, sterilization and UHT treatment of these products and the tests for satisfactory processing they are required to pass (SI 1974, No. 1356 in Scotland).

*Addition of preservatives to milk*
No preservative may be added to milk.

See The Milk and Dairies (Preservatives) Regulations, 1962, SI No. 1531.

*Ice-cream*
Ice-cream is a food which can support the growth of pathogens if its temperature, or that of its constituents before it is made, is allowed to reach and remain at or above 7·2°C. It is heat treated by either pasteurization or sterilization to ensure the destruction of pathogens which could be present. The regulations in the three regions of the UK require the same treatment of ice-cream mixes and of ice-cream.

See The Ice-cream (Heat Treatment, etc.) Regulations 1959, SI No. 734 and as amended by SI 1962, No. 1287 and SI 1963, No. 1083. Alternatively, see The Ice-cream (Scotland) Regulations, 1960, SI No. 2108 and as amended by SI 1963, No. 1101, and SI 1970, No. 1285, or The Ice-cream and other Frozen Confections Regulations (Northern Ireland), 1968, SR and O No. 13.

These Regulations require that the ingredients used in the manufacture of ice-cream are pasteurized by one of three specified methods, or sterilized, and thereafter kept at a low temperature until the freezing process begins. It is an offence to sell ice-cream which has not been so treated.

The mixture of ice-cream ingredients is not to be kept for more than 1 hour at a temperature in excess of 7·2°C before being sterilized or pasteurized.

---

* All Regulations concerned with the designations of milk give details of the exact manner in which the various tests must be carried out.
† In Scotland: SI 1965, No. 253 and its amendments SI 1966, No. 1573 and SI 1975, No. 1997.

Table 38    *Categories of untreated milk in Scotland*

| Category | After milking the milk must immediately be cooled to a temperature not in excess of: | Maximum permissible count after cooling but before dispatch to the retailers Bacteria/ml | Coliforms |
|---|---|---|---|
| Premium | 7·2°C | 15,000 | None in 1/100 ml |
| Certified | 10°C | 30,000 | None in 1/10 ml |
| Standard | 15·5°C | 50,000 | None in 1/100 ml |
| Tuberculin tested | 15·5°C | 200,000 | None in 1/100 ml |

*Source*: SI 1965, No. 253 and its amendments SI 1966, No. 1573 and SI 1975 No. 1997

Table 39    *Statutory tests for milk designations in the UK*

| Type of milk | England and Wales | Scotland | Northern Ireland |
|---|---|---|---|
| Untreated (variously described in the different regions) | Must pass the MB reduction test | All categories are limited in the maximum number of organisms/ml which they may contain; and in their coliform content – see Table 38 | Must pass the MB reduction test |
| Pasteurized | Must pass the phosphatase test | Must pass the phosphatase test Must contain no coliforms in 1/10 ml | Must pass the phosphatase test Must contain no coliforms in 1/10 ml |
| Sterilized milk | Must satisfy the turbidity test | Must satisfy the turbidity test | |
| UHT | Must satisfy the colony count test: 1 loopful of sample after incubation under the conditions of the test to produce not more than 10 colonies | Must satisfy the colony count test: milk after incubation under the conditions of the test shall not contain more than 1000 bacteria per ml | |

To achieve *pasteurization* the mixture is to be raised to not less than:

65·6°C and kept at this temperature for 30 minutes or

71·1°C and kept at this temperature for 10 minutes or

79·4°C and kept at this temperature for 15 seconds

To achieve *sterilization* the mixture is to be raised to:

149°C and kept at this temperature for 2 seconds

After heat treatment the mixture is to be reduced to a temperature of not more than 7·2°C within 1·5 hours and kept at this temperature until the freezing process is begun.

It should not be offered for sale unless since it was frozen it has been kept at a temperature not exceeding −2·2°C. If it has exceeded this temperature it should be retreated by sterilization or pasteurization before sale.

### Liquid egg

Liquid egg is any mixture of yolk or albumen, other than reconstituted dried egg, and includes any such mixture which is frozen, chilled or otherwise preserved.

Some eggs contain or are contaminated with salmonellae. These eggs can infect the bulk of liquid egg, and so pasteurization of liquid egg is now required by law to prevent salmonellosis from this source.

The Regulations in the three regions of the UK require the same method of treatment of the bulk egg.

See The Liquid Egg (Pasteurization) Regulations, 1963, SI No. 1503, or The Liquid Egg (Pasteurization) (Scotland) Regulations, 1963, SI No. 1591, or The Liquid Egg (Pasteurization) (Northern Ireland) Regulations, 1963, SR and O, No. 244.

No person may use liquid egg in food intended for sale unless it has been treated in the following manner.

The egg must be pasteurized by being held at a temperature of not less than 64·4°C for at least 2·5 minutes, and immediately cooled to a temperature below 3·3°C. Samples so treated must pass the α-amylase test as described in the Regulations.

Exception is made for liquid egg removed from the shell on the premises and used forthwith, or kept at a temperature not exceeding 10°C and used within 24 hours.

## Control of addition of substances to foods

This section only considers those which have a direct effect on the microbiology of foods.

### Addition of preservatives

#### General

There are many chemicals which can be added to foods – to improve their colour, to improve their flavour or nutritional quality, to stabilize the food, improve its flowing qualities, or otherwise aid its mass-production, or to discourage the growth of micro-organisms within the food. All these substances are known as *additives*, and there is much legislation controlling their use. This section is, however only concerned with the legislation which controls the addition of *preservatives* to foods – substances which repress the growth of micro-organisms.

The Regulations in the three regions of the UK specify the same requirements.

See The Preservatives in Food Regulations, 1979, SI No. 752 and its amendments SI 1980, No. 931. Alternatively, see (Scotland) SI 1971, No. 1073, amended by SI 1980 No. 1232 or (Northern Ireland) SR and O 1980, No. 28.

These Regulations state that only specified foods may contain *permitted* preservative. Schedules at the end of the Regulations list the types of food which may contain preservative, and the nature of the preservative and its concentration in each case.

A permitted preservative is defined as:

'Any substance which is capable of inhibiting, retarding or arresting the growth of micro-organisms, or any deterioration of food due to micro-organisms, or of masking the evidence of any such deterioration but *does not include*: any permitted antioxidant, artificial sweetener, colouring agent, emulsifier, improving agent, miscellaneous additive solvent, stabilizer, vinegar, any soluble carbohydrate sweetening matter, potable spirits or wines, herbs, spices, flavouring agents, sodium chloride, smoking substances.

*Permitted perservatives include*:

Sulphur dioxide (or Na, or K metabisulphite; Ca sulphite; or Na, or Ca hydrogen sulphite)
Propionic acid (or Na, Ca or K propionate)
Sodium nitrate (or K nitrate)
Sodium nitrite (or K nitrite)
Benzoic acid (or Na, K, or Ca benzoate)
Methyl-4-hydroxybenzoate (or Na salt)
Ethyl-4-hydroxybenzoate (or Na salt)
Propyl-4-hydroxybenzoate (or Na salt)
Biphenyl
2-hydroxybiphenyl
Sodium biphenyl-2-yl-oxide
Sorbic acid (or Na, or K, or Ca sorbate)
Nisin
Hexamine
2-(thiazole-4-yl) benzimidazol

*Milk*
No preservative may be added to milk (see the 'Milk' section, page 116).

*Addition of other substances*

*Meat*
Before 1964 certain substances, now prohibited by the Regulations below, were sometimes added to meat by butchers in order to preserve the fresh appearance of the meat. However, because they can mask deterioration of the meat and certainly in the case of nicotinic acid produce unpleasant side effects in consumers, they have now been banned.

See The Meat (Treatment) Regulations, 1964, SI No. 19, or The Meat (Treatment) (Scotland) Regulations, 1964, SI No. 44 or The Meat (Treatment) Regulations (Northern Ireland), 1964 SR and O No. 6.

These Regulations prohibit the addition of ascorbic, erythorbic and nicotinic acid, nicotinamide or any salt or derivative to raw and unprocessed meat. The Regulations also prohibit the sale of such meat.

## Control of techniques used in processing

### Irradiation
If food is exposed to ionizing irradiation it is possible there is a risk that, as a result of irradiation, the chemical composition of the food may be altered and it may then contain substances harmful to human beings. There is also the risk that if irradiation is used regularly to reduce the microbial flora of food that the process may select the more resistant organisms which would then have the chance to become dominant either in the food or processing environment.

At the present moment because of the lack of sufficient knowledge of the effects of irradiating food the uses to which irradiation are put in the food industry are very limited. The Regulations in force in England and Wales, Scotland and Northern Ireland are in agreement.

See The Food (Control of Irradiation) Regulations, 1967, SI No. 385 and SI 1972, No. 205, The Food (Control of Irradiation) (Scotland) Regulations, 1967, SI No. 388 and SI 1972, No. 307, or The Food (Control of Irradiation) Regulations (Northern Ireland), 1967, SR and O No. 51 and SR and O 1969, No. 226.

No person shall in the preparation of any food (intended for human consumption) subject it to ionizing radiation.

The Regulations also prohibit the sale of any food which has been irradiated. The Regulations do not prohibit the subjection of food to not more than 50 rads of ionizing radiation where the energy of radiation delivered does not exceed 5 million electron volts. Certain other exceptions are made.

*Ionizing radiation* is radiation capable of producing ions – certain X-rays, γ-rays, (these two being of the greatest interest in the food industry), electrons, positrons, protons, α-particles or heavy particles.

### Sterilization
The heat processes of sterilization and pas-

teurization have already been mentioned in connection with milk, ice-cream and liquid egg – and in these contexts details of the processes are given in the relevant Regulations. In addition all *knacker meat* and meat imported otherwise than for human consumption must be sterilized before it enters the chain of distribution.

Here the term 'sterilized' means treated by boiling, or by steam under pressure until every piece of meat is cooked throughout; or dry rendered, digested or solvent processed into technical tallows, greases, glues, feeding meals or fertilizers.

Exception from the Regulations is made for provisions to zoos and menageries, meat supplied to medical and veterinary schools for use in diagnostic or instructional purposes, and also for meat for manufacturing chemists for the manufacture of pharmaceutical products.

The weakness of these Regulations is that they do not stipulate chemical or other tests to ascertain that the sterilization process is satisfactory.

The Regulations for the UK are the same in intent: The Meat (Sterilization) Regulations, 1969, SI No. 871. In 1981 views were invited by MAFF from interested organizations – that is those substantially affected by the Regulations – as to how they are working.

## Other regulations

### Meat inspection

Regulations exist which provide for the inspection of carcasses and the passing of them as fit for human consumption, and their marking to that effect. Indications of unfitness for human consumption are listed in a schedule at the end of the Regulations: if on inspection of a carcass a meat inspector is satisfied that an animal was suffering from any of the listed conditions, the inspector regards the *whole carcass* as unfit for human consumption. The conditions listed include: anthrax, foot and mouth disease, acute septic mastitis, swine erysipelas, swine fever, tetanus, generalized tuberculosis and tuberculosis with emaciation.

See The Meat Inspection Regulations, 1963, SI No. 1229, amended by SI 1965, No. 1487; SI 1966, No. 915; SI 1975, No. 656; SI 1975, No. 882 and SI 1981, No. 996. Parts of these Regulations are also amended by The Fresh Meat Export (Hygiene and Inspection) Regulations, 1981, SI No. 454. Alternatively, see (Scotland) SI 1961, No. 242, amended by SI 1963, No. 1231; SI 1976, No. 874 and SI 1979, No. 1563, or The Meat Inspection Regulations (Northern Ireland) 1967, SR and O No. 8. (*Note*: See also page 189 for Regulations for slaughterhouses.)

### Imported food

No person may import food intended for sale for human consumption which has been examined by a competent authority and found at the time of the examination to be unfit for human consumption, unsound, unwholesome or otherwise unfit.

With certain exceptions meat and meat products cannot be imported without a recognized official certificate from the country of origin guaranteeing that the meat or meat product has been prepared in accordance with satisfactory standards of hygiene.

See The Imported Food Regulations, 1968 SI No. 97, amended by SI 1973, No. 1351; SI 1979, No. 1426) and SI 1981, No. 1085. Alternatively, see (Scotland) SI 1968, No. 1181 amended by SI 1973, No. 1471; SI 1979, No. 1537); and SI 1981, No. 1035, or (Northern Ireland) SR and O 1968, No. 98 amended by SR and O 1973, No. 1350; SR and O 1979, No. 1427; and SR and O 1981, No. 1084.

*Imported milk* can only be received by persons registered with the local port health authority. It must be received in such a condition that it contains no more than 105 bacteria per ml and be free from tubercle bacilli. See The Public Health (Imported Milk) Regulations, 1926, SR and O No. 820.

### Shellfish

Shellfish thrive in waters where the organic content is high because shellfish feed by filtering out the organic matter. There is risk to human health because they can therefore contain pathogenic organisms picked up from sewage in the water. Regulations were introduced to protect the public from this risk.

If a Medical Officer of Health (MOH) is in possession of information that anyone is or has recently suffered from a disease attributable to shellfish he or she may take steps to find out from which layings (the foreshore, bed or other place from which shellfish are taken) the shellfish are derived, if he or she feels that there is liable to be danger to public health. The MOH must report to the local authorities providing them with bacteriological and other reports.

The local authority is empowered to make orders to prohibit the distribution of shellfish for sale for human consumption if it thinks fit. This prohibition may be temporary or permanent in the interest of public heatlh.

See The Public Health (Shellfish) Regulations, 1934, SR and O No. 1342 and as amended by SI 1948, No. 1120.

### Poultry

The Poultry Meat (Hygiene) Regulations, 1976, SI No. 1029 prescribe conditions for the production, cutting up and storage of poultry meat intended for sale for human consumption. The Regulations are amended by SI 1979, No. 693 and SI 1981, No. 1168. They implement EEC directives 78/50 and 77/27.

They also provide for:

1 The *health marking* of poultry meat; describe the hygiene conditions required in the slaughterhouse including *building facilities*; and require a minimum temperature for water used in disinfecting implements (+82°C).
2 The *hygiene requirements* as to staff, premises and equipment, slaughter and evisceration, immersion chilling; ante-post mortem inspection; conditions concerned with poultry meat intended for cutting and boning, wrapping and transport.

In Scotland the SI 1976, No. 1221, amended

by SI 1979, No. 768 apply and in Northern Ireland the SR and O 1979, No. 261 applies.

### International carriage

The International Carriage of Perishable Foodstuffs Regulations, 1979, SI 1979, No. 415 prescribe the temperature conditions for the international carriage of perishable foodstuffs and the thermal efficiency of transport equipment.

The highest temperature permitted in the load for frozen and non-frozen foods during carriage and unloading is as follows:

### Frozen foods

| | |
|---|---|
| Ice-cream; frozen or quick(deep) frozen concentrated fruit juices | −20° C |
| Frozen or quick(deep) frozen fish | −18° C |
| Butter and frozen fats | −14° C |
| Frozen red offal and egg yolks | −12° C |
| Frozen poultry and game | −10° C |
| Frozen meat | −10° C |
| All other frozen foods | −10° C |

### Non-frozen foods

| | |
|---|---|
| Meat other than red offal | +7° C |
| Butter | +6° C |
| Industrial milk | +6° C |
| Meat | +6° C |
| Game | +4° C |
| Milk (raw or pasteurized for immediate consumption) | +4° C |
| Poultry and rabbits | +4° C |
| Red offal | +3° C |
| Fish (must be in ice) | +2° C |

### Water

The Water Act, 1973 together with The Water Act, 1945 embody the legislation which controls the supply and quality of water in England and Wales.

Section 11 of the 1973 Act provides: 'That it shall be the duty of a water authority to supply water within their area'. The authorities have a duty to supply water for domestic purposes. This water must be wholesome. In this context the authority is empowered to add substances to counteract undesirable (pathogenic) organisms, and to add or subtract components to render the water free from suspended or dissolved matter. In other words, to treat it in such a way that they can provide *potable* water.

Water authorities must also supply water for non-domestic purposes on terms which are reasonable to owners or occupiers of premises who request such supply. This supply need not be given if it would impair the authorities' ability to meet its existing obligations, or its probable future requirements to supply water for domestic purposes.

## Waste disposal

These matters are broadly covered by the Control of Pollution Act, 1974.

For more information on water and waste disposal, see Charles Webster's *Environmental Health Law* (Sweet and Maxwell 1981).

## Food hygiene codes of practice

*Code number*
3  Hygiene in the retail fish trade
4  The hygienic transport and handling of fish
5  Poultry dressing and packing
6  Hygiene in the bakery trade and industry
7  Hygiene in the operation of coin-operated food vending machines
8  Hygiene in the meat trades (replaces codes 1 and 2)
9  Hygiene in microwave cooking

## Food Legislation in the EEC

The European Economic Community (EEC) is a group of countries co-operating under the terms of the Treaty of Rome. Among other things the Treaty requires that the Community shall remove obstacles to the free movement of goods by working towards the approximation of the laws of the Member States – a process commonly referred to as 'harmonization' of legislation. In this way legal barriers to trade between the Member States are removed. The means for this harmonization is set out in a section of the Treaty devoted to the 'Approximation of Laws' (Article 100).

*Directives* are negotiated by Member States, and when agreement is reached, which often takes several years, they are issued under Article 100. They are binding only as regards the ends to be achieved, and leave the means to the national laws in the individual Member States.

The legal changes required in the UK when Directives on food law are implemented are executed by making secondary legislation under the Food and Drugs Act, 1955, the European Communities Act, 1971 and the corresponding Scottish Acts.

Community law is also made in the form of *Regulations* issued under Article 43 of the Treaty. This Article governs the establishment and implementation of the Common Agricultural Policy and although it is used in the context of setting quality standards for agricultural products, its main purpose is to lubricate the day to day operation of the agricultural regimes. Regulations can be made extremely quickly (sometimes the whole process only takes a matter of days). Microbiological standards rarely crop up in food law at present (an example is the Natural Mineral Water Directive, 80/777), but they could, in theory, be made either in the form of Directives or Regulations, depending on the circumstances.

# Appendix 2 Pests encountered in the food industry

The activities of rodents, insects, birds and other pests damage food, crops, buildings and other structures. The precise extent of the damage which they cause is unknown as there is very little reliable information and often high figures are quoted which perhaps tend to exaggerate the position. In addition to causing physical damage, however, they are also responsible for the transmission of harmful organisms including food poisoning bacteria.

The pests encountered in the food industry may be studied under the following headings:

1 Rats and mice
2 Insects
3 Birds

## Rats and mice

These pests consume stored food and contaminate it with hairs from their fur and with their urine and faeces making it unfit for human consumption. They present a hazard to health in that they can be carriers of the organisms causing Weil's disease, salmonella food poisoning, bubonic plague, trichinosis and other diseases. They can also endanger property as a result of their gnawing activities. Two species of rat are found in this country – the common rat, and the ship rat. Of the two the common rat is the more widespread, the ship rat generally being confined to areas around ports.

### The common rat (Rattus norvegicus)
The common rat is also known as the brown rat or the sewer rat.

Figure 77 *The common rat*

### Appearance
It is usually brownish-grey with a lighter coloured belly. It has thick short furry ears and a blunt snout. It weighs about 330 grams and is about 425 mm in length including the thick tail which is shorter than the head and body combined.

### Habitat
It is very widespread but tends to be found in places such as food premises, farms, warehouses, rubbish tips, sewers and drains where food is plentiful.

### Habits
The common rat can climb, burrow and swim. It nests both indoors and outdoors and tends to be fixed in its habits. Its droppings are spindle-shaped arranged in groups or occasionally they may be scattered.

### The ship rat (Rattus rattus)
The black rat or roof rat are two other names of the ship rat.

## Appearance

The first noticeable difference between the two species of rat is that the ship rat is much smaller and lighter than the common rat, having an overall length of 375 mm and weighing approximately 250 grams. The ears are large, thin and furless and the snout is pointed while the tail is usually longer than the head and body. The animal's colour varies from black to grey and brown with a much lighter underside.

## Habitat

It was originally introduced into this country from ships which had come from Mediterranean and Eastern ports, and is still found infesting the holds of ships and the dockland areas of ports and from these places it has spread to many large towns.

## Habits

Unlike the common rat, it does not burrow and is seldom found in sewers or outdoors. It climbs very well and nests indoors often in roofs. Both species of rat are very wary of new objects and the ship rat is very erratic in its habits. Its droppings have been described as being sausage-shaped and are scattered.

Both types of rat breed profusely. The common rat is capable of producing from five to eight litters each year, each litter containing up to twelve young rats which mature in approximately 3 months being themselves then capable of breeding fresh litters. Rats possess incisor teeth which grow continuously throughout life and which must be worn down by constant gnawing. The enamel on the outer surface of these teeth is very hard and they can gnaw through almost any kind of material. As a result, they frequently cause fires and flooding in buildings by biting through electric cables, plugs and water pipes. Damage from burrowing sometimes causes subsidence in roads and pavements which in turn affects underground services as well as the surface of roads and pavements.

Figure 78   *The house mouse*

## The house mouse (*Mus musculus*)

### Appearance

It is very tiny, weighing less than 28 grams and is brownish-grey in colour with large ears, a long slender tail and a pointed snout.

### Habitat

It is very widespread and is found in buildings where there is a food supply and shelter.

### Habits

Like the common rat, the mouse burrows and climbs. It has very erratic feeding habits, but unlike rats, is not suspicious of new objects. Its droppings are widely scattered and look like tea leaves. The mouse, like the rat, also gnaws.

### Signs of rodent infestation

There are numerous signs which indicate the presence of rodents. They can be detected by the greasy smears which they leave on brushing their coats against pipes, walls, girders etc., and also the 'loop smears' as a result of running under floor joists along beams. Other signs which may be observed are gnawing marks and holes, feet marks in dust or grain, runways and droppings. Droppings are one of the most important signs of infestation as from them the rodent officer can tell whether the infestation is recent, the kind of rodent, the size of the colony and if there are young rodents present.

### Principles of rodent control

There are several characteristics of rodents that

must be borne in mind when attempting their prevention and extermination. They are nocturnal animals which need shelter and a supply of food and water. The common rat is rather conservative in its habits whereas the ship rat, like the house mouse, tends to be erratic.

### Prevention of infestation

Therefore in considering the prevention of rodents in food storage and preparation areas the aim should be to limit:

1 The food and water
2 Means of access
3 Shelter available to them

### Food and water

All food and water should be kept in rodent-proofed containers. Edible refuse should be placed in bins with tightly-fitting lids, if possible, and if not, the containers should be housed in 'proofed' buildings which are capable of being easily cleaned. Rubbish of any kind should not be allowed to accumulate as it may offer shelter and food.

### Means of access

Strict control of entry to rats and mice must be ensured and although this is best considered during the design of the premises, remedial work to existing buildings should, if necessary, be carried out. Entry and its prevention may be effected in several ways:

Via drains and sewers. Maintenance checks should ensure that all water seals to interceptor traps are effective; there are no 'breaks' in the drain or sewer runs; all manhole and grid covers are whole and close fitting; all WC pans, basins, sinks and sanitary fittings are kept in good condition.

Via small openings in the structure of the building. A small mouse is able to pass through a 10 mm wire mesh while rats may jump up to nearly 1 m in height. All openings therefore which occur lower than 1 m from the ground should be limited to no more than 6 mm or so. Good modern building techniques however are generally free from such openings. Care should

be taken, however, where pipes, pipe sleeves, joists and girders, etc. pass through the exterior skin of the building to ensure that these items are tightly built in or carry sleeves or collars of expanded metal, mesh or wire cloth. All doors, windows, glazed partitions or panels which reach to floor level should be close fitting.

By climbing. This may take place via the inside or outside of vertical ventilation, or rain water pipes, or via rough stone or brickwork walls. The rodents then gain access to the premises by means of ill-fitting or broken doors, airbricks, windows and ventilators. Pipes may be protected by the use of wire balloons internally and metal collars externally, while cement rendering or painting of the wall with gloss paint prevent climbing by rodents; and should these finishes be in a light colour, tell-tale smears and tracks would show rodent presence at an early stage.

By gnawing. Although rats have been known to gnaw many apparently unlikely materials, some such as timber are particularly vulnerable to this form of attack. Care should be taken in the choice of materials used in the lower parts of the building fabric, particularly if the materials are non-traditional such as plastics, to ensure that they are either rodent-proofed or suitably protected. Timber doors, for instance, should be protected by 'kicking plates' in metal fitted to the outside of the door and including the door frame. Foundations of 750 mm or over in depth, which is normal good building practice, is generally sufficient to deter burrowing below the building structure.

### Shelter

The common rat will live outside but, given access, will feed indoors. Internal harbourages should be avoided wherever possible and those that are unavoidable should be inspected regularly. Typical places of this kind are spaces behind skirting boards, between upper floors and ceilings, hollow partitions, ducts, conduits, insulation mats and panels. Suspended ground floors with through ventilation are now virtually obsolete forms of construction but may be found in existing premises. These floors are particularly vulnerable to rodent attack as air bricks are often broken and rarely maintained allowing access to a sheltered underfloor space.

### Eradication of rats and mice

Once rodent infestation is detected plans should be put into operation immediately to exterminate it. The three commonest methods of killing rats and mice are the following:

1  Trapping
2  Fumigation
3  Rodenticides

### Trapping

Trapping is suitable for eliminating very small infestations and as a temporary means of preventing re-invasion of premises cleared of rodents. Break-back traps of the treadle type should be used and, for rats, placed at right angles along known runways. An attractive bait should be utilized as rodents have definite food preferences. Bait may include butter, cake or chocolate, depending on the food sources available.

As rodents are highly suspicious of new objects traps should be left unset for a few days and should not be moved from place to place but extra ones put down if required in other locations. One of the advantages of using traps, especially where there is food on the premises, is that no poison is involved. Mice, unlike rats, show little wariness of new objects and so the traps can be placed and immediately set. It is usual to set a greater number of traps for mice than for rats as they feed from many stations.

### Fumigation

Fumigants are not selective in their action and are dangerous to man, animals and domestic pets as well as the pests against whom they are aimed, and therefore have to be used by skilled operators. The fumigants most commonly used are methyl bromide and hydrogen cyanide and occasionally ethylene oxide and others. Proprietory powders which give off hydrogen cyanide when they come into contact with damp air or soil may be used to gas colonies of rats outdoors, especially in their burrows. These powders when not in use should be stored in airtight tins with tightly-fitting lids. Cyanide dust, however, must not be used in or near buildings in wet or windy weather. The use of hydrogen cyanide is governed by regulations which must be strictly adhered to.

### Rodenticides

The poisons used against rats may be classified under the following headings – acute (single dose) and chronic (multiple dose).

*Acute poisons.* These are usually single dose or direct poisons and up until 1950 they were the only kind used to control rodents. These poisons were used initially without prebaiting but the results obtained were not always good. As rats exhibit bait shyness more successful results were obtained by prebaiting, that is, placing unpoisoned baits down for the first 2 or 3 days to condition the rat to feed. When using these acute poisons they must be kept away from other animals and the dose must be sufficient to kill the rats quickly, for if they experience any unpleasant effects they will stop feeding from these poisoned sources. Only three acute poisons can be recommended for use in food premises – zinc phosphide, nor bromide and alphachloralose. Zinc phosphide can be used for both rats and mice but it must be used with extreme caution and by skilled operators. Nor bromide is not harmful to other animals and is used specifically for rats, being particularly useful for killing Warfarin resistant rats. Alphachloralose may be used to control mice indoors.

*Chronic poisons.* Great advances were made in the field of rodenticides when it was realized that certain anticoagulants which were used to treat various vascular conditions in humans could be used to destroy rodent colonies. The most widely used rodenticide which works on this principle is Warfarin which has many advantages including the fact that it can be used by unskilled operators and is harmless to man and animals in the doses given. It is usually administered as food in dried baits (oatmeal) and acts as an anticoagulant which, if eaten in sufficient quantities, will cause haemorrhaging and death. Its effectiveness relies on the fact that rats feed from the same situations daily and will thus build up to a lethal dose in about a week. The ship rat has more erratic feeding habits than the common rat and therefore in order for Warfarin to be effective against it more baiting points will be required, but as it is smaller than the common rat, less bait. As rats are highly suspicious by nature the poison must be slow acting but not violent so that they do not associate the death of the colony with the poisoned food, which must also be more attractive to them than the food stored on the premises. On the whole, Warfarin tends to be more effective against rats than mice as mice tend to have far more erratic feeding habits, sampling up to 100 feeding stations in 1 day. Therefore in attempting their control with Warfarin a large number of bait points are required.

Baits should be placed in containers which are designed in such a way that the scattering of bait during feeding is limited and laid down in areas known to be used by rodents. The bait is usually dry but sometimes oily baits are used and in sewers a preservative is added to prevent mould growth. Warfarin is also available in liquid form but in order to be effective in this kind of preparation all other sources of liquid must be eliminated. Anticoagulants are also available in the form of dusts which the animals pick up on their fur and which they ingest when they groom themselves. When anticoagulants are used in this form care must be taken that they are not laid in positions where they can blow on to food as they contain more poison than the dry baits.

Rodents which have shown resistance to Warfarin have been detected in several parts of the country, which include Powys and several Scottish and English counties. This resistance to Warfarin seems to be due to the mutation of a single gene. Two other anticoagulants in current use are coumatetralyl and chlorophacinone. The latter may be used as an alternative to Warfarin but it is not always successful against rats which have become resistant to Warfarin.

When resistance is shown to both Warfarin and chlorophacinone, coumatetralyl may prove to be effective. The most efficient treatment for mice is canary seed which contains 0·1 per cent calciferol and 0·025 per cent Warfarin. Mice resistant to Warfarin have been successfully killed using bait dosed with alphachloralose which causes death as a result of hypothermia, that is, a lowering of the body temperature. Alphachloralose, however, cannot be used in heated buildings (temperatures over 18°C).

At all times rodenticides should be kept out of the reach of children and animals and away from foods, and stored in clearly labelled containers. Any eradication programme should be followed by a re-infestation protection programme.

## Insects

Insects, like rats and mice, are capable of damaging stored foods as well as presenting a danger to health. They are capable of transferring infectious diseases and food poisoning organisms from their sources to food. For these reasons they must be kept out of premises concerned with the production, preparation and serving of food.

### *Insects which present a health hazard*

#### *Flies*

The flies most commonly found infesting premises are the housefly (*Musca domestica*), the lesser housefly and the blowfly. The eggs of the housefly are laid in a food supply and develop into *larvae* which feed on the surrounding matter. The larvae or maggots, which are white, remain in the larval stage for about 5 days after which they develop into *pupae* from which the adults emerge (see Figure 79). The whole life cycle takes approximately 10 days in warm conditions. Breeding takes place during the spring and summer and usually ceases in October, but can continue through the winter if flies are in warm surroundings. Hazards to health arise from contamination of food from the dirty body parts of the flies, from their regurgitated food and from their faeces. As the adult flies can only take in food in liquid form they eject saliva on to the food and then suck up

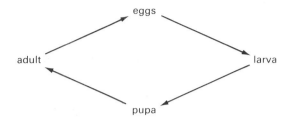
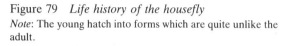

**Figure 79** *Life history of the housefly*
*Note*: The young hatch into forms which are quite unlike the adult.

the liquid mixture. This action is accompanied by regurgitation of previous food which may have been infected.

*Prevention*
The most important control method for flies is to remove the breeding sites. If this fails other methods have to be employed to prevent the entry of flies into premises and to control them once they have entered.

In order to keep flies out of premises the points of entry have to be safeguarded. Windows screened with wire gauze, hanging curtains of beads or plastic strips in doorways, air curtains and double doors at access points are some of the ways in which their entry into premises can be prevented.

*Control*
The conditions within the premises should be made as unattractive as possible to all insects. This may be done by keeping the establishments scrupulously clean. All food scraps should be wrapped, removed and placed in bins with tightly-fitting lids where flies are unable to gain access. All refuse bins should be emptied regularly and not be allowed to overflow. Regular cleaning prevents them becoming breeding grounds for flies. Machinery should either be sealed to the floor so that no food can accumulate beneath it or should be sufficiently raised to ensure that adequate cleaning underneath is possible.

*Insecticides*. Before dealing with insecticides in relation to flies it may help to identify the ones

discussed if they are first classified. The ones used to control insects in food premises belong to one of the following groups:

1  *Insecticides of plant origin*. The only important insecticide which belongs to this group is *pyrethrum* obtained from the flower heads of plants which are produced for this purpose in certain African countries. Pyrethrum supplies are limited but this is overcome by incorporating pyrethrum with a synergist, that is, a chemical which increases its activity. The pyrethrum in these preparations is used in lower concentrations than when it is used alone, and the synergist commonly used is piperonyl butoxide.

2  *Organophosphorus insecticides*. When first prepared these insecticides were very toxic not only to insects but also to birds and mammals. Newer members have been developed which are harmless except in very high doses, examples being *malathion, dichlorvos, bromophos* and *fenitrothion*.

3  *Carbamate insecticides*. These have similar effects to the previous group, two examples being *carbaryl* (sevin) and *propoxur* (arprocarb), the latter being particularly effective against cockroaches.

4  *Organochlorine insecticides*. Certain restrictions have been placed on these insecticides because of their persistent qualities. 'The Advisory Committee on pesticides and other toxic chemicals' report in 1969 recommended that restrictions be placed on the use of insecticides which break down slowly. The report confined itself to certain persistent organochlorines which included *dieldrin, DDT* and *gamma-BHC* (lindane) although the latter is not as persistent as other organochlorines. Dieldrin is an insecticide only to be used by skilled operators against cockroaches and ants. The Committee considered that the recommendations covering the various formulations of dieldrin were adequate but they could not justify its use in insecticidal smoke generators except in places away from people and food, for example heating ducts and roof spaces. DDT

is produced in a variety of forms but the Committee recommended that its use in food storage practice should be discontinued and be restricted to insecticidal smoke generators for use in empty stowages or stowages where there is protection for any food present.

Gamma-BHC was not considered to be as dangerous as other organochlorines so its use in food premises is permitted. Restrictions however were recommended for its use in thermal vaporizers where a person would be exposed to its effect for more than the normal working day. It was also recommended that thermal vaporizers should not be used indiscriminately however harmless the insecticide, but only used where there was actual infestation. The restrictions discussed apply to food storage practice. With reference to use in the home, the Committee stated that dieldrin in small retail packs should be discontinued, the use of DDT should stop and also the use of gamma-BHC in thermal vaporizers.

Insecticides for controlling flies should only be used as a last resort when all other methods have failed. They are available in many forms which include sprays, aerosols, lacquers and powders. For domestic purposes insecticides in the form of aerosols in pressurized containers are very useful against flies and other insects on the wing. Pyrethrum or pyrethrum plus piperonyl butoxide are 'knock down' insecticides which are effective against most flying insects and are considered to be among the most suitable sprays for use in food premises. They do not however have any residual effect. When using an insecticide as an aerosol or spray care should be taken that neither sprayed material nor the dead insects drop into or on to the food. Insecticides may be used in thermal vaporizers which release a measured amount of insecticide at regular intervals although as already stated these should not be used indiscriminately even if the insecticide is harmless to man. Dichlorvos is an insecticide which is available in aerosol cans and as resin strips. The effects of the aerosol preparation disappear quickly in the same way as pyrethrum unless it is mixed with a more persistent insecticide. The resin strips which slowly release dichlorvos vapour are very effective against flying insects and may give protection for up to 3 months. Although dichlorvos has been reported to have adverse effects on man and animals its use in the form of resin strips and aerosols is considered to be safe.

Lampshades and other surfaces can be treated with an insecticidal spray which gives protection for several months and examples of suitable residual insecticides which may be used for this purpose are bromophos (1 per cent) and fenitrothion. Swarms of flies in places such as roof spaces can be dealt with by using smoke generators containing an insecticide such as dieldrin, while malathion has been successfully used on refuse tips to control flies which have become resistant to DDT.

*Electrocution* is another way to control flies. Several devices for suspension from walls and ceilings have been designed which electrocute flies, the insects being attracted to ultra-violet light and hanging objects. The flies do not fall to the surface below but are collected in a tray which should be emptied regularly.

### Cockroaches

The two types of cockroach which are most commonly encountered in the kitchens of food premises, shops, factories and other similar buildings in this country are the German cockroach (*Blatella germanica*) and the Oriental cockroach (*Blatta orientalis*). Two other kinds of cockroach, the American and Australian, may also be found in this country, having been introduced in consignments from abroad.

The *German cockroach*, also called the steam fly, is light brown in colour and about 14 mm in length. It is unable to fly but can run up vertical surfaces and is found in basements, kitchens, bakehouses, food factories and other similar places which provide food, moisture and warmth. During their lifespan the females produce four to eight egg cases each containing approximately 40 eggs. The egg cases are carried at the rear of the insect for approximately 1 month before the eggs hatch into *nymphs*.

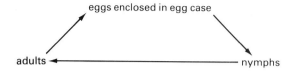

Figure 80   *Life history of the cockroach*
*Note*: There is no larval or pupal stage. Nymphs are very similar to adults, but they are devoid of wings and are much smaller.

Under suitable conditions of warmth and moisture they develop through several moults into adults within 2 months (see Figure 80).

The *Oriental cockroach* is dark brown in colour, about twice the size of the German variety. It is sometimes called the black beetle despite the fact that it is not a true beetle and, like the German cockroach, it does not fly. It is found in places similar to those favoured by the German variety, notably warm, damp situations. The female of the Oriental cockroach during her adult lifespan (5 months) will lay up to five egg cases each containing approximately 16 eggs. The egg case is deposited in a concealed location from which the eggs hatch into nymphs after about two months depending on the temperature; the warmer the surroundings the more rapid the metamorphosis, though this process may take up to 2 years.

Both type of cockroach are nocturnal and during the day may be found in warm dark areas in cracks and crevices behind sinks, boilers and pipes. Their flattened bodies enable them to go through very narrow openings. Like most pests they are found where there is a supply of food, water and warmth. They have an objectionable odour which persists on any surface or food with which they come in contact.

### Prevention
The important signs of infestation which should be recognized are regurgitated food, coloured droppings, moult cases and the smell. Cockroaches are nocturnal, only coming out during the day if disturbed, so that, unless these signs of infestations are noted, their presence could go undetected. Pyrethrum spray may be used in the daytime to flush out the cockroaches to see the extent of the infestation. All scraps of food should be removed to reduce any incentive to the cockroach to stay on the premises. All cracks and crevices which offer harbourages to them should be filled in and the building kept in a good state of repair. All equipment should be sited in such a way that it is easily possible to clean behind and underneath it, or it should be sealed to the floor.

### Control
The best method is to use insecticides. The insecticides used against cockroaches are usually in the form of sprays, powders and lacquers. Sprays may be supplied as suspensions or emulsions in water or as a solution in oil. Pyrethrum plus piperonyl butoxide as a solution in oil is suitable against cockroaches. Oil penetrates the waxy cuticle allowing insecticidal material to come into contact with the insect tissues. The sprays used against flies are not very effective against cockroaches, heavier preparations being needed.

Residual treatment against flying and crawling insects is best achieved by applying liquid insecticides to surfaces over which insects crawl. Porous surfaces are best treated with water-based emulsions or water dispersable powder while non-porous ones are treated with oil-based formulations or resins. Fortnightly application is recommended until control is achieved.

Lacquer, which is a varnish-like liquid, can be applied to surfaces where it sets to a hard transparent coating which is not noticeable. It crystallizes and any insect walking over it picks up crystals on its legs and body parts which when absorbed through the cuticle or ingested when the insect cleans its limbs are enough to kill it. Lacquers are applied in such a way as to make a barrier between the insect and its supply of food. Dust may cover over the lacquer but it can be reactivated by wiping with a damp cloth. It can be applied in bands to skirtings, door frames and table legs. Arprocarb and dieldrin are applied in this way, the former also being effective against cockroaches which have become dieldrin resistant.

In places which are difficult to clean such as heating ducts dusting powders can be used. Dieldrin dusting powder has a very long life but should only be used by trained operators.

Before resorting to any insecticide other methods of control should be applied first, particularly good hygienic practices. No insecticide however harmless should be allowed to come into contact with food.

### Ants

The two types common in Britain are the pharaoh's ant (*Monomorium pharaonis*) and the common black ant (*Lasium niger*).

The *pharaoh's ant* nests within the fabric of the building, being adapted to living in warm conditions. It is often found in modern buildings which are centrally heated and where there is a food supply. The *black ant* makes its nest outdoors around the foundations of the house and usually enters in search of food, being particularly troublesome in the summer.

### Prevention

All particles of food should be removed from the floor thus depriving the insect of its food supply.

### Control

Ants can be controlled in various ways, for example using insecticidal lacquer to paint places with which they are likely to come in contact. They can also be controlled by using chlorocene. Foraging ants eat it and take it back to the nest, the result of which is the eventual destruction of the nest. Its use must be controlled in food factories because of the danger of contaminating food. Spraying of ant tracks with pyrethrum can also give some measure of protection while dieldrin dusting powder is effective against ants as well as cockroaches.

### Wasps

The wasp (of which there are seven species in this country) is a pest of the food industry. The German wasp (*Vespula germanica*) is the one of greatest economic significance. It is attracted to sweet foods such as processed fruits and sugar and causes great inconvenience because of its ability to sting.

### Prevention

To prevent their entry into the factory the same precautions are applicable as for houseflies, that is, double doors and air curtains. They can be killed in flight with the same insecticides that are used to kill flies.

### Control

The best method of control is to seek out the nests and destroy them. To control the nests, attractive baits containing, for example, the carbamate insecticide sevin may be laid down. Wasps feed on it, carry it back to the nest where it is distributed and thus the colony is killed.

### Insects which infest stored food

Insects, as a result of their activities in stored food, do a considerable amount of damage. There are a number of insects which are likely to be encountered in stored foods and so in this section a few of the commoner ones will be described.

Two main kinds of insect are encountered, these being *beetles* and *moths*.

### Beetles

These insects, like moths, have four stages in their life history. The larvae range from active mealworms to sluggish grubs. The mouthparts of both the larvae and adults are adapted for biting and both stages are capable of damaging foodstuffs. The hind wings are membraneous and are protected by the front pair of wings which are hardened for this purpose. The beetles and their larvae produce powdery droppings which give the area in which they are found a dusty appearance.

### The grain weevil (*Sitophilus granarius*)

The grain weevil varies in length from 2·5 to 4·7 mm and is dark brown in colour with an elongated snout. Its hind wings are atrophied and therefore it cannot fly. The female lays her eggs in holes bored by her in the grain and which she then seals with a gelatinous fluid. The larva lives within the grain and is a small legless grub. It changes into the pupa while still within the grain. Once the adult stage is reached it eats its way out of the grain and the life cycle is complete. The adult lives for 7 to 8 months. One of the chief signs of infestation by the grain weevil is holes in the grain. Heavy infestations by these insects frequently causes heating of the grain, a condition which favours the survival of insects. The grain weevil attacks wheat, oats, barley, maize, rye, rice and flour.

The rice weevil, which has a similar history to the grain weevil, also attacks the same range of grains.

*The confused flour beetle (Tribolium confusum)*
The confused flour beetle varies from 3·5 to 4 mm in length and is a dark reddish brown in colour. The female lays about 500 eggs which are sticky to the touch and tend to adhere to the food in which they are laid. The yellow larvae hatch out in about 1 week and reach their full size in 22 days. They are very active in the larval stage and when fully grown they change into pupae. When the adult emerges from the pupa it is light brown in colour but it soon changes to a darker shade. The rate of breeding is dependent on the temperature but they seem unable to survive cold weather. One important sign of infestation by these beetles is the sour smell of the flour. They have a preference for ground grain products rather than sound grain and so *Tribolium* infestations are regarded as secondary.

*The Australian spider beetle (Ptinus tectus)*
The adult Australian spider beetle is dark brown in colour and is about 3 mm in length. It has six legs whereas a true spider has eight. About 120 eggs are laid in the food in which it is breeding, and the small larvae hatch out in about 1 week or less. They burrow through the food, eventually finding their way through the sacking, and lie on the inner surface where they spin cocoons within which the larvae change into the pupae and from which the adults emerge. The life cycle may be complete in just over a month or it may take longer if the temperature is low. There are many signs of infestation which include the presence of cocoons on the inner surface of the sacks and the flattened droppings which are said to resemble wood shavings. The beetle is often found as a pest in flour stored in warehouses and it can withstand low temperatures and a dry atmosphere.

*Saw-toothed grain beetle (Oryzaephilus suri-namensis)*
These insects do not appear to breed at below 20° C. They are small, being between 2·5 and 3·5 mm in length. They are so named because of the serrated edges of the thorax. The female is stated to lay up to 300 eggs, the white slim larvae hatching out in approximately 3 to 17 days. The larval period under favourable conditions may be as short as 2 weeks while the pupal stage lasts 10 days. The beetle is brown in colour and is very active, having a flattened body which enables it to penetrate packaged food. It is a pest in mills and can infect flour and cereal products, and can also be present in machinery.

There are many other beetles which may be encountered, examples being the biscuit beetle and the lesser grain borer.

### Moths
The moth body is covered with scales and the mouthparts are adapted for sucking up juices. The larval stage is the only one capable of damaging food stuffs and the mouthparts of this stage are especially adapted for biting. The larvae are typical caterpillars.

*The warehouse moth or cacao moth (Ephestia elutella)*
The adult moth varies in length from 12 to 14 mm and is grey in colour with a pair of lighter coloured bands on the forewings. The female lays her eggs on the surface of grain but the larva penetrates the grain and consumes the germ. The greyish-white larva is fully grown after 60 to 70 days and throughout the whole of its development it spins a silky thread from its mouthparts which it leaves as a webbing. The larvae leave the grain and crawl up walls into

cracks and crevices where they go into a resting stage. A small number may turn into pupae and they may emerge as adults in October. The majority, however, remain in the larval stage until May of the following year when they pupate. Signs of infestation are the sour smell and the webbing and droppings of the larvae. It usually infests grain, pulses, cocoa, dried fruits and nuts and is most vulnerable in the larval stage when it is moving about.

### The Mediterranean flour moth or the mill moth (Ephestia kühniella)

This moth has long been known as a pest in granaries and flour mills. It rarely attacks commodities other than flour but in flour mills it spoils far more than it consumes. The adult is about 13 mm in length and greyish in colour with black markings on the fore wings. The female lays up to 350 eggs from which the pink larvae emerge. These spin a web which causes clumping of the grain and flour, a consequence of which is the blocking of milling machinery. The fully grown larva spins a cocoon in the machinery, sacking and similar places where it pupates. The adult appears after about 3 weeks. It can develop in flour of low moisture content and also in whole grain. When there is heavy infestation there is 'souring' of the grain and flour.

### Mites

In addition to insects, mites are also encountered as pests in stored foods. They differ from insects in being much smaller and the body is only separated into two parts whereas the insect body is divided into three. Another distinguishing feature is that the insect head bears antennae while the mite head does not.

### The flour mite (Acarus farinae also known as Tyroglyphus farinae)

The adult mite is about 0·5 mm in length. It thrives in moist conditions and in fact needs a supply of moisture. It can however survive low temperatures, ceasing feeding a few degrees above freezing. The flour mite damages wheat by burrowing into and eating the germ within it.

Its presence can be detected by the 'minty' smell it imparts to commodities. When flour infested with mites has been used for making bread the finished product has a sour taste and may not rise adequately.

### Prevention

The absence of cracks, crevices, grooves, ledges, etc., in the construction of the building, fitments and machinery help to prevent infestations in food premises by not giving the insects suitable harbourages. Where these things are unavoidable, however, they should be cleaned out by brushing followed by some form of suction.

Any spillages of flour, grain and other commodities should be swept up and removed with any other debris. Different consignments should be kept separate and stored away from walls to allow for cleaning and a strict rotation of stocks implemented. All consignments should be inspected on arrival for the presence of insect pests and later at monthly intervals. When containers such as sacks are empty they should be removed from the premises and stored away from the food storage areas.

### Control

Insect pests for the most part are unable to stand extremes of temperature and use is made of this fact in their control in food premises. The life cycles of most pests of stored foods are slowed down or stopped at low temperatures. Grain may be cooled by drawing in atmospheric air through it via an installed system of air ducts. Flour mills may be sterilized by heating, the temperature required being between 49 and 55° C and maintained at this level for 10 to 12 hours, the operation being carried out during non-working hours.

Insect pests may also be killed by mechanical devices such as the entoleter. In this machine the flour is thrown against the revolving discs which kills the pests at all stages. The entoleter is now a standard piece of equipment in most flour mills.

*Using chemicals* is another way to control insect pests. There are various forms of chemical control, one of the important ones being fumigation. This involves the use of chemicals which gas the pests and, despite initial strong objections to their use, flour mill fumigation is now almost an annual event. The most suitable fumigant in use is methyl bromide although in some instances some materials fumigated by it acquire a strong smell. Methyl bromide fumigation of products such as grain or cocoa beans is carried out under gas-proof sheets or in fumigation chambers by highly skilled professional operators. Because they are dangerous there are regulations concerning the use of fumigants which must be strictly adhered to. Other fumigants which may be used are ethylene oxide and hydrogen cyanide.

Insecticides may be used to combat insect pests of stored foods as well as the ones which present a danger to health. The formulations in which the insecticides are used depend on the locations to be treated. Pyrethrum plus piperonyl butoxide may be used as a spray to treat vulnerable places where insects may be found and it is particularly effective when applied to these places after cleaning. It may be used to treat warehouses during the summer months by 'mist spraying' and is particularly effective against the warehouse moth, but the treatment has to be repeated at frequent intervals as it has no residual effect. Pyrethrum can also be used in a smoke generator but this is not as effective as the various fumigants mentioned, as it does not penetrate the stored products but rather tends to settle on surfaces. Bromophos an organophosphorus insecticide may be mixed with grain up to a level of 12 ppm and gives protection against stored pests such as the Mediterranean flour moth. Malathion, another insecticide of the same group, is particularly effective in farm grain stores for controlling the saw toothed grain beetle.

When applying control measures against insect pests in the food industry, the characteristics and habits of the particular pest must be borne in mind when deciding control measures. It is therefore important that the pests are identified before attempting their control as this will save time and money. Precautionary measures, however, should always be operated.

**Birds**

A third group of pests are birds such as sparrows and pigeons. They can contaminate food with their droppings which can carry food poisoning organisms, mainly salmonellae. Measures that can be taken to prevent their entry into premises include repairing ventilators, broken windows and similar places. Positioning of polypropylene netting to exposed roof members and other potential roosting sites will protect goods awaiting despatch from droppings and feathers. Bird repellant strips can be placed on window sills and other perching sites, the number of which should be minimized. The use of poisoned or stupefying baits against birds is prohibited except under licence.

# Index

Numbers in italics refer to pages on which figures and tables occur.